Pigs

Keeping a Small-Scale Herd

BY ARIE B. MCFARLEN, PHD

An Imprint of BowTie Press®
A Division of BowTie, Inc.
Irvine, California

Barbara Kimmel, *Editor in Chief*
Sarah Coleman, *Consulting Editor*
Joe Bernier, *Book Design and Layout*
Indexed by Melody Englund

Reprint Staff:
Vice President, Chief Content Officer: June Kikuchi
Vice President, Kennel Club Books: Andrew DePrisco
Production Coordinators: Tracy Burns, Jessica Jaensch
BowTie Press: Jennifer Calvert, Amy Deputato, Lindsay Hanks
Karen Julian, Elizabeth L. McCaughey, Roger Sipe, Jarelle S. Stein
Cindy Kassebaum, *Cover design*

Library of Congress Cataloging-in-Publication Data

McFarlen, Arie B.
Pigs : keeping a small-scale herd for pleasure and profit / by Arie B. McFarlen.
p. cm. — (Hobby farms)
Includes bibliographical references and index.
ISBN 978-1-933958-18-7
1. Swine. I. Title. II. Series.

SF395.M36 2008
636.4—dc22

2007020565

BowTie Press®
A Division of BowTie, Inc.
3 Burroughs
Irvine, California 92618

Printed and bound in China
13 12 11 10 4 5 6 7 8 9 10

Pigs

This work is dedicated to my grandparents, Clifford and Margery Mann, for inspiring me to make farming my life, and to George and Esmeralda, who know what being a pig is all about.

Table of Contents

INTRODUCTION

Why Pigs?

Pigs are one of the oldest domesticated animals and one of the most valuable to humans. Today's pigs are descendants of wild boars first domesticated in Asia and Europe several thousand years ago, when human societies shifted from being nomadic hunting and gathering based to settlement and agriculture based. Traditionally, the pig served as a primary food source in civilizations around the world, and no part was wasted: the pig has been a source of oil for cooking and lubrication, leather, brush bristles, and fertilizer, among other things.

Today, pigs are still an important commodity. Modern husbandry produces leaner, specialized swine breeds for cured products such as ham, sausage, and bacon and for fresh cuts such as chops and spareribs. Pigs have important medical uses as well: pig insulin and heart valves have successfully been used to treat human diseases for decades. And in some places, small breeds of the sociable pig have become popular pets.

Farmers interested in raising pigs for profit can do so easily as well as produce meat for their own freezers. Farm-raised pork is appealing to many people who are interested in knowing where their food comes from, the conditions in which the hogs were raised, and the nature and quality of the pigs' diet. Consumers—including gourmet home cooks, professional chefs, and ethnic and specialty markets—appreciate the ability to purchase directly from the farm, and farmers can potentially realize higher profits by selling their healthier and more flavorful pork.

Pigs can also be utilized to improve your property or complement your other livestock production. A farmer can take advantage of a pig's natural habit of rooting to clear brushy, weedy, or rough areas of a property, enabling and preparing the area to be reseeded or planted with a valuable crop. Pigs are extremely efficient utilizers of feedstuff and can fatten quickly on the wasted morsels of other animals.

CHAPTER ONE

Pigs: A Primer

Understanding the basics of pig evolution, biology, and behavior can provide valuable insight into selecting the right breed for your farm and caring for your new pigs. Here's a brief history of pigs and an overview of pig types, breeds, and traits.

THE PIG'S PLACE IN HISTORY

Prehistoric drawings of wild boars can be found in Spain's famed cave of Altamira, a dwelling of Cro-Magnon humans some 30,000 years ago. The artwork of ancient civilizations that followed depicted pigs in all sorts of settings, even in scenes with royalty and deities, suggesting that pigs have been familiar, useful animals throughout human history.

The domestication of wild pigs may have occurred first in central and eastern Asia. According to zooarchaeologist Richard Redding at the University of Michigan, 11,500-year-old pig bones have been recovered at Hallan Cemi, in southeastern Turkey. Further research indicates that these pigs were domesticated, predating the cultivation of cereal grains. Anthropologists also believe the Neolithic people of the Peiligang culture in China (7000 to 5000 BCE) raised millet and pigs as their primary food sources. Evidence of this cultivation has been recovered by an excavation site located at Jiahu, led by archeologist Shu Shi. And the earliest known book on raising pigs was recorded in 3468 BCE by Emperor Fo Hi of China.

Additionally, the pig has been an important food source in Europe for thousands of years, where it was both independently domesticated and introduced through trade and migration from the Far East. Pork products and lard were used

Large numbers of pigs have traditionally been fed in small areas, thus maximizing production, as seen in this antique photo of a small-scale hog lot.

to sustain Roman armies, and owning pig herds was a sign of wealth in Anglo-Saxon Europe.

The pig has been both praised and maligned throughout history. Many a pig has been referred to as gluttonous, filthy, or fierce (or all three). No doubt some pigs do possess these qualities, but in general, pigs are not that way at all. Not until the early Middle Ages did cultural prejudices lead Europeans to view pigs as filthy or lowly creatures. Pigs acquired this undeserved reputation largely through teachings of the church, the most powerful institution in the medieval Western world. Various churches taught that demons took frightening, disgusting forms with the physical characteristics and habits of animals, especially pigs.

Despite cultural attitudes that lowered their status, pigs were raised in large numbers by Europeans, who eventually took their pigs to the New World lands they explored and conquered during the Renaissance. There, a pig's adaptability to varying foliage worked against domestication. Many pigs turned loose in the New World quickly adapted to new habitats and became feral hogs, widely roaming the American colonies by the 1600s. Over the centuries, immigrants from all lands brought their native pig breeds with them, and the pig population of North America became a hodgepodge of mixed breeds.

Pig populations spread across the United States as settlers moved westward. By the 1840s, a growing proportion of American pigs were raised on the fertile soil and plentiful corn of the Midwest and Great Plains. With the advent of the refrigerated railcar, pork

A pair of backyard pigs are curious, waiting to see if they will get a treat from onlookers. Pigs such as these would have been seen throughout the midwestern states as early as the 1830s.

Pigs as Good Luck

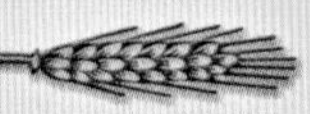

In many countries, pigs are considered to be good luck. In Ireland, for example, a peppermint-flavored candy Lucky Pig is wrapped in a velvet sack and given as a present. A tiny hammer is used to smash the pig while making a New Year's wish, and the candy is shared by all.

Lucky Pigs, called *Sparschweinchen,* are given as a New Year's present in German-speaking countries.

The nickname for pigs in early America was the "mortgage lifter," as raising pigs brought property owners good fortune of another kind: no other meat requires so little investment for such quick and continuous returns, and profits from the sale of pigs at market could easily pay a property owner's mortgage.

Pigs have been considered good luck in many ethnic traditions. A wish of good luck was sent to friends and family, such as the one seen in this antique greeting card.

production in the United States became solidly concentrated in the Midwest.

Through industrialization and the growth of cities during the 1800s and 1900s, the backyard pig and small-scale butchers became a relic of bygone days. Beginning in the early 1900s, the general population gradually gave up producing meat in favor of supermarket convenience and the cost benefits of mass-produced products. According to http://www. Pork.org, farms raising fewer than 1,000 meat hogs per year represent only 1 percent of the total pork market in the United States. Large-scale operations selling 500,000 or more hogs per year represent over 40 percent of the total pork market. The United States Department of Agriculture (USDA) states that in 1980, more than 600,000 family and commercial farms were producing pork in the United States. As of 2000, fewer than 100,000 pork-producing facilities remained. This means that fewer and larger farms are producing the bulk of the market hogs, creating niche market opportunities for the hobby farmer or small producer.

Pigs in Mythology

In Celtic mythology, the boar represents fertility, wealth, courage, and warrior spirit. Many Celtic works of art depict warriors standing with a boar, and many crests of ancient Celtic families bear the symbol of a boar or boar's head.

In classical Greek mythology, a sow suckled Zeus, chief god of the Greeks. Swine were sacred to Demeter, goddess of the earth's fertility and mother of Persephone, queen of the underworld. In autumn, during the rites of Thesmophoria, devotees of Demeter would throw pigs, bread, and pine branches into a cavern as sacrifices. Later, they would return to see if the deity had accepted this offering by examining the condition of any pig carcasses that remained. This cult was later absorbed by that of the Roman goddess of grain, Ceres. Swine were also sacrificed to the Roman deities Hercules and Venus by people seeking relief from illness.

In Homer's epic poem *The Odyssey,* Greek hero Odysseus encounters the sorceress Circe, who turns his crew into swine. With the aid of Hermes, who provides him with an antidote to Circe's swine brew, Odysseus is able to resist Circe's magic and persuade her to set his men free. Circe then changes the swine back into men of even greater stature and handsomeness.

The pig appears throughout world mythology: The Beast of Cornwall, for example, is described in British medieval literature as a boar. Varahi is a boar-faced Hindu goddess believed to protect Newari (Nepalese) temples, buildings, and livestock. And the Hindu god Vishnu, incarnated as a boar, rescued the earth by balancing it on his tusks after it had been hurled to the bottom of the sea by an evil demon.

Pigs: The Breeds

There truly is a hog breed for every person and husbandry method. Eight major hog breeds are raised in the United States, plus several heritage, pet, and minor breeds. All breeds differ in growth rate, litter size, mature body size, time required to reach market weight, and grazing ability.

All major pig breeds and most minor breeds in the United States have their own official breed registry. These registries maintain pedigrees and statistical data. Registries are the best place to start when researching pig breeds. They offer a wealth of information about the breed itself, including the history of the breed, the breed standard, and any health or specific traits within the breed. Registries also provide breeders' lists and contact information, facilitating your research on your chosen pig breed. Some registries even offer Web pages to advertise your pigs, as well as want ads and announcements.

A pig breed not only can be selected for color, size, and personality, but pigs also can be selected based on statistical data regarding their expected performance. Swine Testing and Genetic Evaluation System, or STAGES, was developed by the National Swine registry to track the most economically significant traits of the breeds as well as make predictions based on genetic potential. STAGES can track litter size and weights, days to market, back fat depth, and intramuscular fat. The information can be used to compare the potential of pigs from the same breed with the production goals of the farm and help swineherds make educated decisions about which pigs to buy.

Pigs will do just about anything to get a meal. This vintage picture shows a very tolerant cow, one who is willing to feed more than her own offspring!

Profiles of Swine Breeds

Pig breeds can be grouped into one of four categories: commercial, endangered, heritage, and pet. The breeds in each category have distinct characteristics and are suited to particular husbandry methods, climates, and production intentions. Carefully evaluating breeds to match your production method will increase your likelihood of success in raising pigs.

A litter of crossbred pigs displaying hybrid vigor—the maximization of different traits among breeds. Hybrids are bred for rapid growth, meat quality, and increased size.

Did You Know?

Breed standard is defined as the ideal or required physical attributes an animal must possess. In pigs, breed standards define such things as ear size or shape, body type, color, mature weight, and sometimes production records. Breed standards for various breeds have been modified over the years, as preferences and desirable traits have been identified.

Commercial Breeds

Commercial hog breeds are commonly seen in large-scale confinement facilities, which are animal feeding operations used primarily to raise pigs to market weight or to farrow sows. These confinement buildings house the pigs throughout their lives. Proponents of confinements claim that this housing method protects the pigs from weather, predators, and disease and allows for easier care, feeding, and management. Commercial hogs are selected for characteristics suited to mass production, such as rapid growth and physical uniformity. Most commercial breeds are hybrids or superhybrids (hybrids crossed with other hybrids) produced in professionally managed, ongoing breeding programs with the primary purpose of improving growth and repro-

Advice from the Farm

Choose the Right Breed

Our experts offer some advice on selecting the pig breed that is best for you.

Selecting Your Pigs

"Select your pigs based on the way in which you want to raise them. Genetics and previous environment determine a lot about how a pig will behave on your farm. It has been difficult to convert confinement hogs to a naturally raised system. They don't know what to do with the bedding, don't know how to behave in groups, and don't really have mothering skills. Pick pigs from farms that are raising them similar to what you will be doing."

—Al Hoefling, Hoefling Family Farms

Know Your Preferences

"Pigs come in all shapes, sizes, and colors. Choosing a pig that matches not only your husbandry methods but also your personal preference is important. Pigs chosen for preference will bring you a sense of pride, and you will take better care of them. While preference is important, try to be objective with your requirements. Many people have color preference or size preference. Choose the breed that matches your personality, but also choose the best animals in that breed."

—Bret Kortie, Maveric Heritage Ranch Co.

Buy Right Before Breeding

"When you purchase breeding stock, you should not be far from breeding, I feel. A lot changes in an animal from 50 pounds to 250 pounds, 'conformationwise'. The top animals should be kept for breeding, not the bottom end, and that is hard to tell when they are really young."

—Josh Wendland, Wendland Farms

These are Durocs, a breed found in many commercial operations. They can range in color from light gold to dark mahogany.

ductive rates. These breeds include, but are not limited to, the Duroc, Landrace, Poland China, and Spotted or Spots. Although considered commercial, these breeds may do well on a pasture-based system. You should select animals from herds already utilizing the husbandry method you intend to use.

Duroc: Durocs are red pigs with drooping ears. They are the second most recorded breed of swine in the United States. Developed from the Jersey Red of New Jersey and the Duroc of New York, Durocs can range from a very light golden to a dark mahogany. Popular for prolificacy (ability to produce large litters) and longevity, Durocs also produce a quality, lean carcass. Most hybrid breeds in the United States include the Duroc, which contributes improved eating quality and rate of gain. A high rate of gain means that the Duroc requires less feed to create a pound of muscle. Breed standard requirements include solid red color; medium length, slightly dished face; and drooping ears.

Landrace: The Landrace is a white pig. Its ears droop and slant forward so the top edges are nearly parallel to the bridge of a straight nose. Landraces, noted for their ability to farrow and raise large litters, are the fourth most recorded breed of swine in the United States. Descended from the Danish Landrace and Large White Hog, the American Landrace also includes Norwegian and Swedish stock. Landraces are known for their length of body, large ham and loin, and ideal amount of finish weight. Landrace sows are prolific; they farrow large piglets and produce an abundant milk supply. These

This aged Duroc boar is still in production. The lifetime service of a boar with high prolificacy can spell profits for the small-scale farmer.

Landrace pigs are used heavily in crossbreeding programs throughout the United States. This Landrace-cross pig displays the qualities sought in large-scale productions, such as fast growth and large muscle.

traits have designated the Landrace breed as "America's Sowherd," and they are heavily promoted in crossbreeding.

Poland China: The Poland China breed had its beginning in the Miami Valley, Butler, and Warren counties in Ohio. It would be difficult to evaluate the exact contribution of any particular breed or type of hog to the Poland China, as it was developed by the crossing and recrossing of many different breeds. Poland Chinas were originally bred for two important characteristics—size and ability to travel—because they were driven on foot to market and in some cases were required to walk nearly 100 miles. Today's Poland China hog is recognized as a big-framed, long-bodied, lean, muscular individual that leads the U.S. pork industry in pounds of hog per sow per year. The Poland Chinas have very quiet dispositions with a rugged constitution. Breed standard requires a black pig with six white points (feet, tail switch, and nose) and flop ears.

Spotted or Spots: The present-day Spots descend from the Spotted hogs, which trace a part of their ancestry to the original Poland China. A later infusion from two Gloucestershire Old Spots boosted the breed with new bloodlines. Spots are good feeders, mature early, are very prolific, and pass these characteristics to their offspring. The breed is well documented: established in 1914, the National Spotted Swine registry has grown to one of the top-ranking purebred breed associations in the United States. Spots have continued to improve in

Gloucestershire Old Spots, as seen here, contributed to the Spots breed of today. Gloucestershires add large litter size, good dispositions, and—of course—the spots to crossbreds.

This is a Mulefoot hog, considered endangered by the ALBC but still a viable production hog. Full bodied, the Mulefoot produces ample amounts of meat and lard.

feed efficiency, rate of gain, and carcass quality, making them popular with both small farmers and commercial swine producers.

Endangered Breeds

Endangered hog breeds are those classified by various breed preservation organizations as being in danger of extinction, either because of low numbers or insufficient genetic diversity to maintain the population. Hog breeds classified as endangered by the American Livestock Breeds Conservancy (ALBC) show a remarkable resistance to parasites and diseases common among commercial pig breeds. These breeds are also highly adaptable to harsh conditions and poor-quality feeds and possess good maternal instincts and long-term fertility. These qualities are highly desirable for the low-input, sustainable agricultural systems practiced on homesteads and hobby farms.

Although these endangered breeds are found in limited numbers, many dedicated conservationists and farmers have maintained and even expanded the populations of these animals in recent years. Homesteaders and niche pork aficionados alike have contributed to the repopulation of these breeds by promoting public awareness of them and creating an end use for the animals.

Even the young pigs of endangered breeds, such as this Mulefoot, are able to forage for themselves and grow well on pasture. Hardiness and the ability to convert roughage to meat is a desirable trait still present in the endangered hog breeds.

The endangered category comprises Choctaws, Gloucestershire Old Spots, Guineas, Herefords, Large Blacks, Mulefoots, Ossabaws, Red Wattles, and Tamworths.

Choctaw: Choctaw hogs are descendants of the pigs brought to the New World by the Spaniards and adopted by the Choctaw Indians of Mississippi. The Choctaw were among several southeastern tribes forcibly relocated to the Indian Territory in Oklahoma, where their hogs were generally established as free-rangers, foraging for all their own food. The Choctaw is a small- to medium-size pig, averaging 120 pounds. Physical characteristics typically include erect ears, wattles, and mulefooted, or single, toes. Generally black, it may have white on the ears, feet, and wattles. The Choctaw is a long-legged pig, able to range widely for food. Today, the bulk of the Choctaw hog population is maintained within the Choctaw Nation in Southeastern Oklahoma, where it remains a vital food source, as it has been for more than 200 years. Choctaw hogs are categorized as critical, with an estimated population of fewer than 200 animals.

Gloucestershire Old Spots: Originating from the Berkeley Valley region in Gloucestershire, England, Gloucestershire Old Spots (GOS) were known there as Orchard pigs. Traditionally, GOS were used to clean up fruit orchards, nut tree stands, and crop residue. GOS are large white pigs with black spots. Their ears are large, are lopped, and cover the entire face to the snout. Mature GOS sows are known for large litter sizes and abundant milk. Full-size GOS will reach 400–600 pounds by the age of two years. GOS are well known as a gentle and sweet-tempered breed. Highly adaptable, they can be raised with a variety of management practices and in varied climates. With proper shelter, GOS will thrive outdoors year-round. They fatten well on a large variety of foods, including fruit, whey, nuts, beets, kale, sweet potatoes, crop residue, and mast (the fallen nuts, fruits, and leaves of trees). GOS produce a fine carcass with top-quality meat for all purposes—chops, roasts, hams, and sausages. Gloucestershire Old Spots are categorized as critical, with an estimated American population of 150 animals. GOS can also be found in the United Kingdom, where they are classified by the Rare Breeds Survival Trust as a minority breed, with fewer than 1,000 animals.

Guinea: A small hog, weighing between 85 and 250 pounds, the Guinea is solid black with medium pricked ears and a straight or very slightly dished snout. It is hairy rather than bristly. The Guinea is a true miniature pig, not a pot-bellied (dwarf) pig, and its body parts are proportionate for its size. Flat backed, Guineas are larger in the shoulder than in the ham. Guineas have been known by several descriptive names such as Yard Pigs, Snake Eaters, Acorn Eaters, and Pineywoods Guineas. Guinea hogs have a sketchy and disputed history but are now considered a unique American breed. Historically, guineas were recorded as having come to the United States via slave ships, or possibly with

Did You Know?

According to the American Livestock Breeds Conservancy, livestock breeds are grouped into the following categories according to the current population numbers:

- **Critical:** Fewer than 200 annual registrations in the United States and estimated global population of fewer than 2,000.
- **Threatened:** Fewer than 1,000 annual registrations in the United States and estimated global population of fewer than 5,000.
- **Watch:** Fewer than 2,500 annual registrations in the United States and estimated global population of fewer than 10,000. Included are breeds that present genetic or numerical concerns or have a limited geographic distribution.
- **Recovering:** Breeds that were once listed in another category and have exceeded Watch category numbers but are still in need of monitoring.
- **Study:** Breeds that are of genetic interest but that either lack definition or lack genetic or historical documentation.

explorers from the Canary Islands. Unfortunately, none of the history can be proved conclusively. According to the ALBC, "Several mysteries confuse the breed's history. The relationship between the historic Red Guinea and the Guinea Hog may be simply the common use of the term 'guinea' to refer to an African origin. 'Guinea' may also refer to the small size of the hogs."

Guineas are known for their friendly disposition and gregarious nature. Guinea hogs are highly adaptable and suitable for sustainable or low-input systems, as they are able to forage and graze well, gaining nicely on grass and weeds. Guineas are good mothers, averaging six piglets per litter. The sows are attentive but not possessive, allowing easy management during farrowing. Meat from the Guineas is fine flavored, though fattier than most other breeds. This fat is desirable for slow-roasting meats and adding rich flavor to other dishes. Guinea hogs are categorized as critical by the ALBC, with fewer than 100 breeding animals in the current population.

Hereford: Hereford enthusiasts claim they raise the "world's most attractive hog," based primarily on the breed's colorful red coat and the white markings on the face, feet, and belly. The Hereford was created from a cross of Duroc-Jersey and Poland China hogs by John C. Schulte of Norway, Iowa, around 1920. They have drooping ears; a wide, slightly dished face; and curly tails. Hereford bodies are even from shoulder to ham with a slight arch to the back. They can be raised on pasture or in semiconfined conditions. Their color and hardiness are well suited to outdoor production, but shade should be provided to protect against sunburn. They grow well on a variety of feeds and do not put on excessive amounts of fat. They like to root and can be useful for tilling. Hereford boars are known for their aggressive breeding habits and are

These Guinea piglets start out weighing less than one pound at birth. They will more than triple their weight during their first week of life.

very prolific; litter sizes average eight to nine piglets. Full-size Herefords range from 600 to 800 pounds at two years of age. Gaining in popularity since the mid-1990s, Hereford hogs now number more than 5,000 in the United States.

Large Black: The Large Black hog is a full-size pig with intensely black pigment in the hair and skin. Originating in the Cornwall and Devon areas of England in the late 1800s, Large Blacks were known locally by their regional names, such as Cornwall and Devon, and as the Lop-eared Black in East Anglia. The founding of the Large Black Pig Society in 1889 led to the exchange of stock between the regions and a uniform name for the breed. Today, Large Blacks can be found in very small numbers in the United Kingdom, Australia, Ireland, Canada, and the United States. Large Blacks have long deep bodies and long straight faces and snouts. The large drooping (or lopped) ears nearly cover the face of the breed, often obstructing its vision. The intensely dark pigmented skin protects the Large Black from sunburn. Fully mature at the age of three, Large Blacks average 500 to 600 pounds. Recognized as hardy and thrifty (that is, they can maintain weight under varying circumstances), Large Blacks were originally raised in rough conditions, left to clean up residue and fallen fruit and nuts from fields, brush, crops, and hardwood forests. Mature sows average ten to thirteen piglets per litter and produce ample milk to feed them. Large Blacks stay in production for eight to nine years, a highly desirable trait for homesteads. Categorized as critical by the ALBC, the Large Black population in the United States is fewer than 150 animals. Similarly, in the United Kingdom, Large Blacks are under the vulnerable status, with fewer than 300 animals noted.

Mulefoot: The American Mulefoot hog is a distinct breed recognized officially since 1908 and recorded since the American Civil War. Breed standard for the Mulefoot, established in the early 1900s, described a medium-size black hog with medium forward ears, soft hairy coat, and hooflike feet. The distinctive feature of the Mulefoot is its single toe, which resembles the feet of equines, instead of the cloven hoof most swine breeds possess. This breed has good, heavy bone structure. A long, straight primitive-looking tail, with a tassel or tuft of hair on the end, similar to tails of wild hogs or rhinos, is desirable.

Mulefoots produce succulent, flavorful meat that is red in color and highly marbled. The breed was once prized as a premium ham hog and lard producer. Mulefoots tolerate both heat and cold very well and can be raised in nearly any

Did You Know?

According to the Guinness Book of Records, the most expensive pig in Britain was Foston Sambo 21, a Gloucestershire Old Spots, which sold at auction in 1994 for 4,000 guineas (about $8,400).

Hardy and strong, the Mulefoot retains much of its primitive features. This boar sports a heavy coat, sturdy back, and the premium hams that make the Mulefoot memorable.

climate. They are excellent foragers and grazers as well as highly efficient users of farm surplus and crop residues. Mulefoot litters average six piglets, but can be as large as twelve. Adult Mulefoots weigh between 400 and 600 pounds by the age of two years.

Classified as critical by the ALBC, the current Mulefoot population numbers fewer than 250 animals. The Mulefoot hog is not available outside the United States.

Ossabaw: The hogs of Ossabaw Island, one of the Sea Islands off the coast of Georgia, are descendants of hogs left by the Spaniards nearly 400 years ago. They have remained a distinct, genetically isolated, feral population ever since. Ossabaws are primarily black with a brown tinge, often with white splotches throughout the body. They have prick ears, long snouts, and heavy coats with thick hackles, similar to a razorback's. Possessing the thrifty gene, Ossabaws are able to put on large amounts of fat during times of ample feed to sustain themselves through periods when food is unavailable. Although not particularly good at grazing or rooting, they are excellent foragers and hunters. Ossabaws will hunt small mammals, birds, and reptiles for food, making them very self-sufficient. Ossabaws average eight piglets per litter and generally wean the same number. Small and agile, Ossabaws do not frequently lie on piglets as is typical with large breeds. Their meat is very tasty, with a firm but not tough texture. Fat is marbled throughout and is deposited as a rind over the ham and shoulders. Ossabaws are relatively large in the shoulder area, yielding more roasts and chops. Ossabaw Island hogs can weigh from 100 to 250 pounds fully grown.

Classified as critical by the ALBC, Ossabaw hogs on the mainland United States number fewer than 100 animals. Importing additional animals from Ossabaw Island is nearly impossible because of state and federal regulations regarding the importation of hogs and the costs associated with the importation.

Red Wattle: The Red Wattle hog is believed to have originated in New Caledonia, an island of Melanesia in the South Pacific colonized by the French. The breed came to North America via New Orleans with French immigrants in the eighteenth century. Although generally red in color, the Red Wattle may have black markings. The head is lean with a straight snout, and the ears are erect. The essential identifying feature is the breed's wattles—fleshy appendages of cartilage that hang like tassels from the lower jaw at the neckline. Each wattle is thumb-size in diameter and grows from one to five inches in length. Red Wattles produce a fine, lean meat that is said to have a unique taste, between that of beef and pork. Red Wattles are particularly good in bio-friendly systems in which the hogs are used to turn compost or root up marginal ground. A Red Wattle sow will typically farrow nine to ten large piglets and produce ample milk to fatten them. Red Wattles can be expected to reach full size of 1,000 to 1,200 pounds by the age of three when fed a balanced diet.

Red Wattles are not found in any country other than the United States. Classified as critical by the ALBC, the Red Wattle population is fewer than 200 animals. A handful of breeders are continuing to raise the Red Wattle hog and register the offspring. Small numbers of pigs are available from this dedicated breeders group.

Tamworth: The first Tamworths were imported to the United States from England in 1882. The recorded history of the breed dates back nearly 100 years earlier. Considered the oldest "unimproved" breed in England, Tamworths have remained relatively unchanged for the past 200 years. The Tamworth is a red-gold colored pig with a straight, fine, abundant coat that is highly resistant to sunburn. Primarily bred as bacon hogs, Tamworths are popular because of their ability to produce a white-fleshed carcass with long sides and decent hams. Tamworth meat is lean and highly flavorful. Aided by their exceptionally long snouts and curious natures, Tams are heavy rooters. This is a very handy trait if a farmer wishes to till rough ground, rustle behind cattle, salvage crops, or raise hogs on marginal ground or in tree stands. Litter size averages about ten piglets, with an exceptionally high weaning rate. Full-size Tams weigh 600 to 800 pounds.

Heritage Breeds

Heritage pig breeds in the United States are those with a history of production, conformation, documented registries, and breed standards. They are often seen as part of a cultural heritage of the region from which they came. Heritage hog breeds have a population large enough to not be considered endangered, and they include the Berkshire, the Chester White, the Hampshire, the Large White, the Middle White, the Saddleback, the Sandy and Black, and the Yorkshire.

Berkshire: Originating in Berkshire, England, the Berkshire is a large black pig with white feet, nose, and tail tip. The Berkshire was imported to New Jersey in 1823, making it the first purebred swine in the United States. Berkshires were heavily crossed with other breeds to bring about improved rate of gain and hardiness. In 1875, the American Berkshire Association was formed to preserve the purebred stock. Berkshire meat is said to be richly flavored, dark red, and well marbled. Breed standard requires short upright ears, medium dished faces, long bodies, and deep sides. Berkshire boars average 500 to 750 pounds, while sows average 450 to 650 pounds.

Chester White: The Chester White breed originated in Chester County, Pennsylvania, through a combination of the Yorkshire, Lincolnshire, and Cumberland pigs from England. The Lincolnshire and Cumberland are now extinct. Three registries were combined to form the Chester White Swine Record Association, with registrations dating back to 1884. Breed standard requires a completely white pig with a slightly dished face; medium floppy ears; and full, thick coat. Chester Whites are known for their superior mothering ability, durability, and soundness. They are preferred by producers and packers for their muscle quality and white skin, which dresses out to a light pink when processed.

Hampshire: This black pig with the distinct white belt is known as "the Mark of a Meat Hog," with leanness, minimal amounts of back fat, and large loin eyes. Originally, only Hampshires were used to produce Smithfield hams. Originating in southern Scotland and northern England, this breed has been highly developed and utilized in the United States. Admired for its prolificacy, vigor, foraging ability, and outstanding carcass qualities, the Hampshire has seen a steady growth in popularity and demand. Hampshire females have gained a reputation among many commercial hogmen as great mothers and have extra longevity in the sow herd (remaining in production for up to 6 years). Hampshires are the third most recorded breed of pigs in the United States, indicating that they are popular on hobby farms as well as in commercial productions.

Large White: The Large White owes its origins to the old Yorkshire breed of England. Large Whites are distinguished by their erect ears and slightly

These Hampshire growers display the breed's characteristic prick ears, strong shoulders, and white belt. Hampshires are a colorful favorite among pig breeders.

dished faces. They are long-bodied with excellent bacon and hams and fine white hair. True to the name, full-size Large Whites average 600 pounds. The Large White is a rugged and hardy breed that can withstand variations in climate and other environmental factors. Although developed as an active and outdoor breed, it does very well in intensive production systems. Its ability to cross with and improve other breeds has given it a leading role in commercial pig production systems around the world. Sows produce large litters and plenty of milk. Their extra height, or length of leg, helps them remain active and have long, useful lives in the breeding pen.

Middle White: Middle White swine originated in the Yorkshire area of England at about the same time and from the same general stock as did the Large White and Small White breeds. In 1852, at a livestock show in Yorkshire, England, a group of pigs with exceptional merits was removed from the Large White group and placed in its own category for judging. The main breed characteristic is the snubbed snout. The white-colored swine are well balanced and meaty. Early maturing, Middle Whites are valued when the object is to produce lightweight marketable pork in a relatively short time. Despite their smaller size, the sows have been found to rear an average of eight pigs per litter. They are good mothers and are known for their docile behavior. The Middle White can make a contribution to cross-breeding programs to improve eating quality.

Saddleback: The Saddleback pig is an amalgamation of the Wessex and Essex breeds. Found mostly in England, the Saddleback is raised as a fresh pork hog. Known for its tolerance to heat, the Saddleback was once raised in the southern United States. Saddlebacks have a white belt similar to the Hampshire's, with large flop ears and a slightly dished face. The Saddleback is the not-so-distant relative of the Hampshire breed and shares several of the Hampshire's production qualities. Saddlebacks are medium-large pigs that produce large litters that grow quickly. Suited for foraging as well as confinement, Saddlebacks can produce well on a variety of feeds.

Sandy and Black: Oxford Sandy and Black pigs are a British breed once thought to be extinct. Efforts are being made to reestablish this breed in its native England, and it was recently recognized as an independent breed by the British Pig Breeders Association. Although wide variations in color occur, the pigs must be sandy (blonde to red) with black blotches (not spots). White-tipped tail, four white feet, and a white blaze are desirable. Ears must be lop or semilop. This breed is renowned for the quality of pork and bacon it produces. Sandy and Blacks are touted as one of the best pig breeds for a first-time pig keeper because of their docile personality, mothering ability, and ease in handling. Breeding stock is currently available only in England, although importations are possible.

Yorkshire: Yorkshires are white in color and have erect ears. They are the most recorded breed of swine in the United States and in Canada. Although known as Yorkshire throughout most of the world, the breed is called the Large White in England, where it originated. Yorkshires must be completely white with erect ears and short snouts. Yorks are very muscular, with a high proportion of lean meat and low back fat. American Yorkshire breeders have led the industry in utilization of the STAGES genetic evaluation program and have amassed the largest database of performance records in the world. In addition to being very sound and durable, Yorks have excellent mothering ability and large litters, and they display more length, scale, and frame than do most commercial breeds.

Pet Breeds

Some people enjoy keeping a pet pig around. Two favored breeds are the Kune Kune and the Pot Bellied.

Kune Kune: The friendly Kune Kune comes from New Zealand, where they have been domesticated since the mid-1800s. *Kune* means "fat and round" in Maori. The Kune Kune's body is round and sturdy, with short legs, an upturned snout, and two tassels hanging from its lower jaw. Kunes are smaller than commercial breeds of pig, usually no heavier than 260 pounds, and they have a very good temperament. Kunes fatten readily on grass and are considered a grazing pig rather than a forager. Kunes produce a quality carcass, although fattier than a commercial pig breed's carcass. Worldwide population of Kune Kunes is now at about 2,000 animals.

Pot Bellied: Developed from the "I" breed of Vietnam in the 1950s, the Pot Bellied pig is a dwarf breed. It is usually black, with loose folded skin and thin hair. Almost comic in appearance, this pig has a face that is deeply wrinkled, a snout that is quite short, an abdomen hanging low to the ground, and disproportionately short limbs. Most people who purchase Pot Bellieds keep them as pets, although—like any pig—the Pot Bellied can be eaten. In comparison with other breeds, it is a much fattier pig, with an average dress out of 36 percent meat and 54 percent fat (the remainder being waste). The average adult Pot Bellied pig weighs between 100 and 250 pounds.

Biology

The biology of a pig is similar to that of humans in many ways. Although pigs and humans have some differences, the pig's anatomy and physiology can be broadly grouped into the following eleven categories:

1. Circulatory
2. Digestive
3. Endocrine
4. Immune
5. Muscular
6. Reproductive
7. Nervous
8. Respiratory
9. Urinary
10. Sensory
11. Skeletal

Pigs such as these can make wonderful pets. They train as easily as a dog and can be taught many tricks, such as walking on a leash and sitting on command.

A group of piglets graze on the farm. Pigs will forage for a large amount of their food if given the opportunity. Pigs can convert grass and other plant material to body growth with ease.

Pigs are omnivores, which means they eat a variety of materials, including plants and other animals. Pigs are not ruminants and do not possess multiple stomachs. Because of their ability to process a large variety of feeds, pigs can be housed on land that would be unsuitable to other livestock. Pigs are able to root, or dig up, food, worms, grubs, and such and so provide a varied diet for themselves.

It is a popular misconception that pigs wallow in mud because they are dirty animals. This is simply not true. Pigs are not equipped with sweat glands, and adequate shade or other cooling measures must be provided for a pig to be comfortable and survive. Mud not only cools the body but also creates a barrier against biting insects.

Pigs have very poor eyesight, which may explain why pigs become easily upset and frightened when they are being moved. If a pig is being asked to move into an unfamiliar space, allow it time to look around and evaluate the situation. Trying to move a pig in to a dark space will cause anxiety, and it will not readily go.

Pigs typically have excellent hearing. Communication among pigs takes place through grunts and squeals. An advanced system of communication can be seen while observing a sow nursing her piglets: sows communicate to the piglets when it is time to eat by certain grunts, and they sing a sort of lullaby to the piglets when they are nursing. An angry or upset pig will perform a series of barking noises to warn the antagonist to back off. When happy, a pig may make a sort of purring or gurgling sound.

The nose knows. Pigs use their sense of smell to find food, to identify each other, and to root. The pig's nose is a highly sensitive, multipurpose tool.

A sense of smell is imperative to a healthy hog. Pigs identify each other through smell as well as use this sense to identify or locate food. The snout is used to gauge distance, to dig up food or make a hole to lie in, for courting mates, and for interacting with other pigs.

A pig eyes the camera. Many pig breeds have poor eyesight, in part because of the wide spacing and small size of the eyes. Pigs rely more on their senses of hearing and smell to survive.

Healthy pigs should have a strong bone structure, which will help support a large amount of weight on small, short legs. The feet of pigs are typically cloven hooves except in the case of a mule-footed, or single-toed, hog. The feet must be sound to support the weight of the pig but also be of good structure to avoid cracked or split toes, abscesses, and arthritic conditions.

Behavior

Pigs are social animals and should be raised in groups. Raising them without any companionship is cruel. If you must raise only one pig at a time, providing it with the companionship of another

Biological Traits of Pigs

Temperature: 101.5 degrees Fahrenheit
Pulse: 70 beats per minute
Respiration: 25–35 breaths per minute
Expected Life Span: 3–15 years
Sexual Maturity: Females 6–8 months, males 4–6 months. Although they are able to reproduce at this early age, females should not be put into breeding service before 6 months of age.
Heat Cycles: Female pigs will cycle about every 21 days, with the actual heat period (time in which she can become pregnant) lasting about 1–2 days at the end of the cycle.
Gestation: 3/3/3; that is, approximately 3 months, 3 weeks, and 3 days, or about 115 days
Color: Pigs can be nearly any color, including sandy, brown, black, white, red, blue (grey), and any combination of colors
Hide: A variety of coats can be seen, from excessively hairy to nearly hairless, as well as curly, straight, smooth, and rough coats

species will suffice. Many pet pigs have their own cat, dog, or goat. Of course, you can also be the pig's companion—if it is allowed to go wherever you go! Pigs raised in groups always eat better and are less destructive, as they are not as prone to boredom. Moving one pig away from the group will cause anxiety for the loner. If possible, move groups of pigs together when taking them to new areas.

Pigs crave companionship and will bond with humans or other animals. Here, a Mulefoot sow expresses her affection for an American bulldog.

Pigs should be raised in groups that remain stable throughout their lifetimes. Introducing a single pig into an established group will jeopardize the pig's safety, as the other pigs will fight with the new addition in an effort to run it off. Reintroduction of pigs to their old family groups should be done through a fence to allow them to safely get reacquainted. Boars should never be moved into existing sow pens, as sows are very territorial and will instigate fights with the boars if they feel they are being invaded upon. When breeding, always take the sow to the boar pen for mating.

Buffy the cat assists the supervision of piglets.

Pigs are highly intelligent animals. According to the National Pork Producers Council, pigs are the fourth smartest animal, preceded by humans, nonhuman primates, and dolphins. This intelligence can be used to the advantage of a farmer or handler. Pigs can be easily trained to come for food with a bell, your voice, or other method. They can be taught tricks, such as opening doors, rolling over, or sitting on command. Pigs are a particular favorite at petting zoos, as they easily develop bonds to humans and want to interact with them.

Pigs are very clean animals. When given the opportunity, individuals and groups of pigs will designate certain areas in their living spaces to be used for different bodily functions. Pigs will choose a separate area for eliminating, sleeping, lounging, and eating. When setting up pens, be sure to always put the feed in the same place, and do not put food in the toilet or sleeping areas. Once the pigs have chosen their area for the toilet, you can place extra bedding there to absorb waste.

Pigs have a variety of personality traits. Getting to know your pigs will

Bill, a Duroc boar, lounges around, waiting for his next assignment. Boars are generally kept by themselves when not in service to the sows.

Like all pigs, these are highly curious. Friendly pigs will come to the fence for a scratch or treat, which adds to the pleasure of taking care of them.

help you manage your herd. Most pigs will settle in once they get to know who you are. They may come running to see you, but they may stand back, away from the fence, when a stranger approaches. Friendly pigs are always more fun to work with, as they can be scratched, petted, and moved much more easily. Do your best to identify the personality of your pigs, and work with them according to their disposition.

When confronted with a stranger or uncertain situation, pigs will generally turn to face you as opposed to running away. If they feel threatened, they will snort, bark, and paw the ground. This is a sign of confrontation and aggression. Try to talk calmly to the pig, and move around it in a slow and nonthreatening way. Do not try to touch an angry pig, as it may try to bite you in an effort to move you away.

Pigs like to root, play, dig, and explore their living areas. Bored pigs will resort to bad behaviors, such as chewing fences, digging, or destroying feed and water dishes. To prevent such behaviors, provide your pig with an enriched environment, perhaps by placing a bowling ball into the pen, providing fresh greens or grass clippings, or introducing companion pigs or animals.

Much research has been done on pig behavior in an effort to reduce stress in handling, both for the pig and the handler. Works by Dr. Temple Grandin are very informative on the topics of stress-free handling of animals and the psychology of their behavior. Through these works, a person can learn to see the world through the eyes of their animals and learn how to anticipate their behaviors. Humane and stress-free handling of your animals will lead to a trusting and more manageable relationship with your pigs.

A pair of pigs enjoy each other's company as they soak in the sun.

CHAPTER TWO

Selecting and Buying the Right Breed of Pig

Don't rush into buying pigs. Although keeping pigs is relatively easy, it will serve you well to learn all you can about the care, feeding, and handling of pigs before you add them to your farm.

Spend some time at the library or bookstore, and study anything you can get your hands on. The Internet offers a lot of information, but be aware that Internet content is not edited for accuracy, and you may be misled. Make attempts to get yourself a mentor, and talk with your local veterinarian, local swine breeding groups, or county extension agent. If possible, visit other farms that raise pigs and ask questions. Most people love to talk about their animals and enjoy sharing their stories. Be considerate of their time, and express your gratitude for their help.

What to Buy

First, you must decide if you are going to raise a couple of pigs for meat in your freezer or if you want to breed and raise pigs for market. If your primary goal is pork for personal use, buying two weanling barrows (castrated males) to feed and butcher is a good way to start. They will provide plenty of meat for a four-person family, plus one pig to sell. This would also be a good learning experience: you will gain invaluable knowledge of handling, feeding, and caring for pigs with little investment or risk. Starting with spring-born weanling pigs will make things even easier, as you will have the pigs in the freezer before winter. This will eliminate the need for their winter housing and other husbandry issues.

If your goal is to breed, raise, and market your own pigs, you should start with the absolute best breeding stock you can get. If the initial cost of the pigs is a deciding factor, start with fewer pigs of higher quality as opposed to more pigs of lesser breeding. Pigs that conform to the breed standard and have proven production records will be the best choice. You will benefit from this general rule in the long and short terms.

Hogs raised on pasture will need some protection from the elements. Portable huts or Quonsets such as these make ideal pasture housing.

Determining Your Production Model

Before selecting your pig breed, decide and understand how you will manage your pigs. Do you intend to raise your pigs on pasture or on forage? Will they reside in the woods, or will you raise them in a confined area such as a barn or a pen?

Most hobby farmers are interested in raising their pigs outdoors, in the most natural environment available at their farm. By allowing your pigs to range on pasture or forage under trees, you are expecting them to glean a large portion of their food supply on their own. If water is set up within your pasture or tree stand, the pigs will require little supplemental care during peak growing seasons and fall harvest time. Pigs that are already being raised in this manner will adapt most readily to this system on your farm. Hogs previously raised in confinement situations may not adapt as well to the change in diet and the level of activity required to forage their own feed.

Sandy areas such as this do not become overly muddy when drained properly. Feeder pigs raised on a sand substrate will remain dry and comfortable.

Confining your hogs gives you the most control over what they eat and where they are located, but it increases your labor and food costs. Confined pigs rely on their keepers for everything they need. Feeding twice daily and assuring an around-the-clock water supply are necessary.

Whether you intend to raise your pigs on pasture or forage, in the woods or in confinement, you should look for

Hogs raised in confined areas must be given all their food, water, and shelter. Even a small area such as this can easily be converted to hog penning.

pigs that come from an environment similar to what you will be providing. This will not guarantee success under your program, but it will at least get you started with pigs that have already proved they can thrive under these conditions.

Choosing the Breed

Let's face it—within certain fixed limitations of your farm and production purposes, you pick your pig breed based on your personal preference. This is the way it should be. If you are enthusiastic about the breed you have chosen, you will take more time finding just the right pigs and will be more diligent in their care once they are yours. That being said, following are some recommendations about choosing the best breed for your farm.

Before shopping for pigs, determine precisely what their purpose will be. Make a list of the qualities or uses you are looking for, and note which ones are essential. Pigs come in all shapes, sizes, personalities, and uses. Some breeds are best suited to particular climates, while

Purchase pigs that are already accustomed to the method in which you will raise them. Starting pigs in a new environment might set them back and inhibit their growth.

others can thrive anywhere. Some pigs are naturally docile, while others are high strung or aggressive.

In addition to being grouped by their official breed name and category, many pig breeds are further classified as one of three types, depending on their primary use: lard, bacon, or pork. Lard pigs have a higher percentage of fat overall and a distinctive shape, known as chuffy. Chuffy pigs look like a sausage, with four short legs and a rounded or arched back. Lard was once the reason for raising pigs, and meat was actually the by-product. Lard was used for liniments, lubricants, and candle oil, but mainly as a dietary source of fat. It was used in cooking for hundreds of years, long before oil was extracted from plant seeds. Lard is still popular today, especially in ethnic cooking and naturally produced medicinals. Nearly all lard pigs fall in the heritage or endangered category.

Bacon hogs have longer legs, longer bodies, and greater depth of rib (larger area from the spine to the middle belly). This depth of rib is the area bacon comes from, and pigs with larger sides produce more bacon. Some heritage and endangered breeds are the bacon type, but most bacon hogs are commercial breeds.

Pork pigs tend to be leaner and grow faster than other breeds do and are used primarily for fresh cuts of pork. Most pork pigs fall in the commercial category, although excellent fresh pork can be had from heritage and endangered breeds.

If you have a clear preference for fresh cuts or cured meats, select a pork

Outdoor hog production usually entailed fencing the hogs out of certain areas, as opposed to fencing them in. This allowed the pigs to gather much of their own food while the vegetable garden or play areas for children were protected from the pigs.

Heritage and Endangered Breeds

Heritage and endangered hog breeds are ideally suited to hobby farms. They can be raised outdoors with minimal feed input and left to forage nuts, fruit, grains, and pasture during many months of the year. They are also highly adaptable and resistant to parasites and diseases, and they produce "old-fashioned" flavorful pork. Pork from heritage and endangered breeds, with the exception of the extra-lean Tamworth, are typically slow growing and contain a higher fat content in the meat. This higher fat makes the meat more tender and juicy, as it bastes the meat while cooking. The full-fat flavor is very similar to the meats produced fifty or more years ago.

The American Livestock Breeds Conservancy (ALBC) was established in 1977 to promote and preserve endangered breeds of livestock. This not-for-profit agency provides information about these swine breeds and offers breeders' lists.

Even the smallest homestead can raise exceptional pork and make a difference in the world by raising one of the endangered breeds. By raising one of the endangered breeds, we not only contribute to the preservation of the species, but we can also assist in public education about such breeds and their benefits in meat production. (See chapter 1 for a snapshot of these breeds.)

or bacon type, respectively. If you have no clear preference, consider a multipurpose breed, a pig that can be used for lard and for fresh and cured meats. Indeed, many breeds are suitable for all purposes, although lard will not be prevalent in the modern commercial breeds. Many hybrid pigs would be suited for multipurpose uses, as well as some of the older British breeds.

Deciding Between Purebred and Crossbred

You must also decide whether you want to raise purebred or crossbred stock. As with the breed choice, this decision depends on what you ultimately want to do with your pigs. Reasons people select purebreds include the desire to breed pigs, raise piglets, and sell breeding

Gwen, seen here relaxing with the author, is a Gloucestershire Old Spots sow. This large breed is categorized as critically endangered and is known for its docile disposition.

stock. In addition to adding pigs to a farm for the benefit of producing their own food, many farmers are concerned with the maintenance and preservation of a species. By raising purebred pigs, you are keeping your chosen breed in production and reducing the likelihood of extinction. Purebred pigs can always be sold as meat, but crossbreds cannot necessarily be sold as breeding animals. If your goal is pork for your own use, consider that different purebred varieties have distinctly flavored meat, and selection should be made per your taste preference. When evaluating breeds for home consumption, purchase some meat from the breeds you are considering to evaluate and compare the flavors.

Just as purebred dogs and cats are sold at premium prices, purebred pigs generally cost more than crossbreds do. Purebred pigs typically are registered or eligible to be registered. Registrations from an appropriate breed association can tell you a lot about your pig: its birth date, parentage, owners, and any championship titles in its pedigree. Associations and registries for various breeds exist in the United States and abroad. Contact these associations and visit their Web pages for a list of breeders and detailed information about the breed you are considering.

Crossbred pigs are of mixed breed parentage and typically cannot be registered. Crossbred pigs certainly have their place on farms across the land. They are generally raised as market hogs, capitalizing on their rapid growth. Crossbreds of two purebred parents have what is called hybrid vigor. This means that the offspring of parents with dissimilar genes have the virtues of both parents to counteract genetic weaknesses—typically, the ability to produce a large, usually fast-growing pig. If your porker is to be used for meat, a crossbreed is a less expensive and perhaps more efficient purchase. If you are raising a pig or two each year for freezer meat and do not intend to go into farrowing and raising piglets yourself, crossbred pigs will satisfy your pork needs.

Where to Buy

Pigs can be purchased from a variety of sources including breeders, small farms, and livestock and exotic animal sales. Ask other farmers if they can make a referral. Many feed stores have bulletin boards listing animals for sale, or you could place your own Pigs Wanted ad. Breed-specific or livestock newspapers and newsletters also frequently post ads for animals for sale.

Buying from Breeders

The best way to buy pigs is directly from the breeder, particularly one who specializes in the breeds you want. Careful screening is important, however: contact a number of breeders and ask questions before you agree to any purchase. Reputable breeders should be willing to describe their operations and affiliations. Ask about their breeding program, how the pigs are raised, what illnesses and diseases they vaccinate for, how long

Sows will protect piglets, even if they are not their own. A squealing piglet will arouse the protective instincts of a sow, whether it is her offspring or another pig's offspring.

they've been raising pigs, and what livestock associations they are affiliated with. Always ask how the pigs are being fed. This will help you to determine whether a particular breeder's stock will do well under your production method. Ask for references and get feedback from other customers to be sure you are dealing with an experienced, knowledgeable breeder.

A breeder truly interested in the welfare of the pigs will screen you, too! Be prepared to tell the breeder what you will be doing with your pigs and how you intend to raise them. If the breeder sees potential problems, he or she should be willing to steer you toward a more appropriate breed.

An experienced breeder should also be willing to help a novice get started by offering specific guidance in raising the pigs you buy, for example, by recommending feed and housing. If you are purchasing purebred stock, the breeder should explain which association your pigs will be registered with. If you are starting with piglets, the breeder should offer information about what age the piglets will be weaned, what vaccines they have received, and the availability of various bloodlines for breeding.

Compare several pigs from the breed you have selected. Look for good conformation; the body and features should be consistent with the breed standard.

Visiting a herd and breeder is the best method of selecting pigs. This gives you the opportunity to view the breeder's herd, management system, and—most important—his or her style of handling the pigs. You get to see firsthand how the pigs behave under their normal living conditions. Observe the feeders, waterers, and other equipment for condition and cleanliness. Are the animals well cared for and the facilities reasonably well maintained? Are the droppings of the animals healthy, or are they pasty or watery? Do the animals appear to be well nourished?

If you are considering one of the heritage breeds, chances are the breeder is not local, and you will have to either travel to buy and pick up your pigs or have them shipped sight unseen. If you buy pigs from a breeder whom you have not met, you have to rely even more heavily on the breeder's reputation. Again, ask for references.

Stockyard and Livestock Sales

You can also buy pigs at stockyard and third-party livestock sales. This is not recommended, however, especially for first-time buyers. Animals sold through such sales are typically culls from a herd, defective in some way or sick. If they were not sick when brought to the sale, they can become sick from exposure to other animals at the sale. Usually, sales do not afford the purchaser the opportunity to talk with the previous owner. You will not be able to get the pigs' history or health records or evaluate their former living conditions.

The Selection

Once you have determined your breed and contacted some breeders, be prepared to evaluate the specific pigs offered for sale before you make your selection. Pigs should be evaluated on conformation, type, disposition, and health. If you purchase proven breeding stock, the production records for potential choices should be available to help you make your decision.

Young adult pigs may be a better purchase, as they will already be displaying their potential for growth and breeding. Older stock will have a shorter production life in your herd, however, so don't choose pigs older than five years of age.

Conformation

Pigs should be true to their type. Although it is difficult to exactly predict a pig's adult size by examining a piglet, some attributes can be evaluated. Piglets should be active, alert, and curious. They should have distinct muscle definition by weaning age, which is approximately eight weeks old, and be well rounded but not fat. Avoid piglets that are runts, skinny, or lethargic.

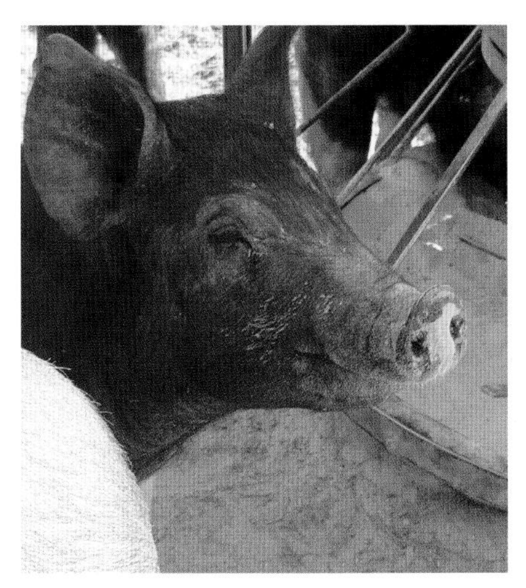

When selecting breeding stock, look for attentive pigs that have good muscling and are not overly fat or skinny.

Breeding-age pigs should be well filled out without being fat. Proven sows should have good teat placement (not pendulous or malformed), with at least ten teats on heritage breeds or twelve teats on commercial breeds. Overly fat sows will not breed well, so avoid such animals. Adult boars should have obvious testicles. If a boar has tusks, consider having him detusked before bringing him home. Tusks can be very dangerous to handlers. (See chapter 5 for complete details on detusking.)

All pigs should have a healthy coat and no bald patches or signs of skin conditions or parasites. Check feet for deformities, injuries (is the pig limping?), or excessively overgrown toes. Tails can be curly or straight (typically depending on the preference of the breeder).

Advice from the Farm

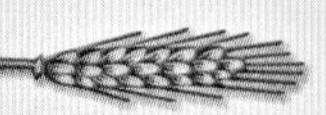

Choose the Right Stock

Suggestions from our experts on selecting pigs for good disposition and health.

Check Disposition

"When choosing pigs for your farm, look for pigs that show an interest in you, come up to see who you are, or are friendly. Do not select pigs that are intensely frightened or do everything they can to get away from you. You will not be able to tell if a weanling boar will be aggressive when mature, but by starting with friendly pigs, you have a better chance of getting ones that will be manageable."

—Al Hoefling, Hoefling Family Farms

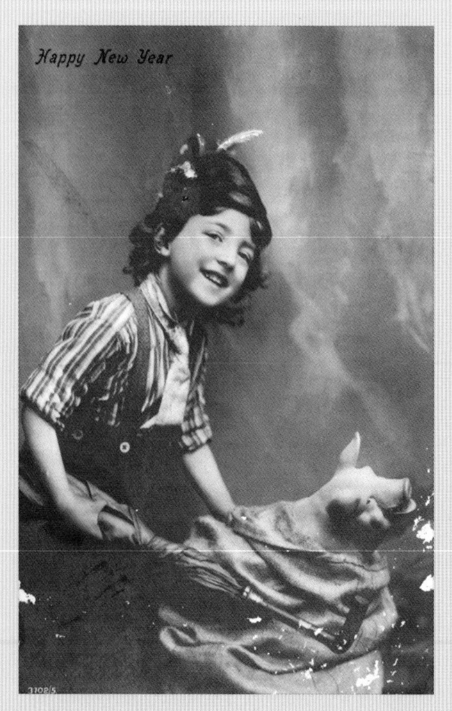

Avoid the Runt

"Don't buy the runt or a sickly pig, thinking that you can rehabilitate it or save it. Runts and sick pigs almost never catch up to the others, and many of them don't survive. Start with the best pig from the litter. It is hard to turn away from a cute piglet or one that you really think is nice looking, but try to look beyond color and markings and be objective in your selection. Choose the piglet with the best conformation, disposition, and health for long-term growth and breeding success."

—Bret Kortie, Maveric Heritage Ranch Co.

Walk This Way

"Whenever you look to buy an animal, watch it walk: it should walk fluidly, not stiffly, and stand on its legs properly. The legs are important. Front legs should be straight and strong; back legs should have good bone. Don't get overexaggerated hams, such as in show pigs, as it restricts the birthing canal. Boars should have the same width at the ham and shoulder."

—Josh Wendland, Wendland Farms

Check Parents

"When buying pigs, look at the parents. Pigs should be free from diseases. Look for healthy ones that are active and interested in what's going on around them."

—R. M. Holliday

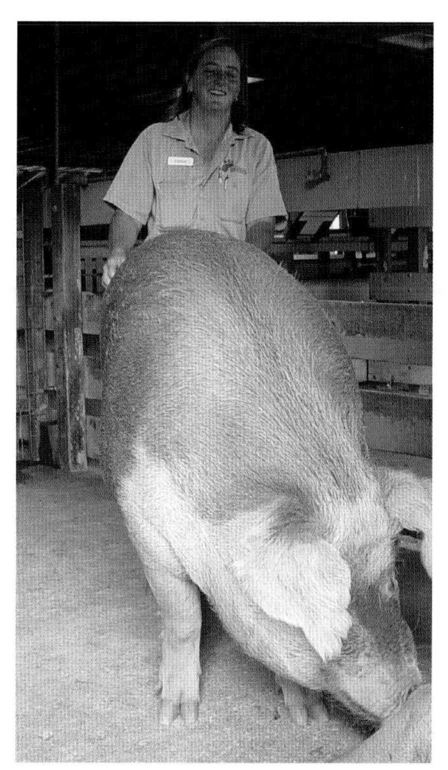

Large breeds have set records for decades. Although she may seem huge, this Hereford sow, shown with her keeper at the Sedgwick County Zoo, is the typical size for her breed.

Health

Never knowingly buy a sick pig. In many cases, careful visual inspection for symptoms will help you rule out a potential purchase. For example, pigs should be free from runny noses and eyes. Avoid pigs that cough, wheeze, or have diarrhea. Pigs with abscesses, lumps, noticeable swelling, or joint problems should also be avoided.

In addition to doing your own inspection, make arrangements for a veterinarian to check out your potential pig purchases and get a full evaluation before you buy. The veterinarian can suggest blood tests that will confirm or deny the existence of diseases in the pigs; can check for parasites through feces and on the skin; and can evaluate the animal's teeth, bone structure, foot health, and general quality. These pigs are your investment, and you have every right to request a professional opinion of the stock. In any case, most states require some sort of health certificate by a licensed veterinarian for pigs transported across state lines. Some states require blood tests to verify that the pigs are free from certain diseases that could be transmitted, such as pseudo-rabies, tuberculosis, or porcine reproductive and respiratory syndrome (PRRS). (See the appendix for more on these and other pig diseases.) Again, your veterinarian can tell you what is required for your state. If a breeder refuses to allow a professional medical evaluation of your potential pigs, look elsewhere for your animals.

Did You Know?

The heaviest domestic pig that ever lived was Big Bill, a Poland-China hog owned by Elias Buford Butler of Jackson, Tennessee. Bill weighed 2,552 pounds and was so large that he dragged his belly on the ground. He measured 5 feet high and 9 feet long. Bill was registered to appear at the Chicago World's Fair in 1933. Unfortunately, he broke a leg and was put down before he could be seen by the general public.

The Sale

When you pick up your pigs, be sure to get a bill of sale identifying the animals you have purchased, their registration status if they are purebreds, the purchase price, and any guarantees the seller has claimed. If your pigs are already registered, ask for the certificates of registry. If it is your responsibility to register your pigs, get a completed and signed application for registry. The application certifies that the piglet is of purebred, registered parents and that the piglet was born of the breeder's pigs.

Some breeders will not accept checks as payment on the day of sale. Prior payment arrangements should be made, or you should be prepared to pay in cash. This is a courtesy appreciated by most breeders.

After the Sale

Once the sale is complete, you can take your pigs home. You can arrange for transportation by an animal carrier or airline, or you can pick up the pigs yourself.

Getting Your Pigs Home

Consider carefully how you want to have your pigs transported to your farm. Although many livestock haulers will deliver pigs for a fee, hiring a carrier that regularly moves pigs can expose your pigs to diseases during transport.

A hog panel, looped around into a teardrop shape and fastened with D-rings, makes a quick and easy hog mover. The pig follows along while the back side of the panel gives them a nudge forward.

Piglets cling together for warmth and comfort. If at all possible, ship two or more piglets together for their own safety and comfort.

If you use a carrier, try to find one whose usual cargo is horses. Horse carriers typically are not exposed to any diseases that can be transmitted to pigs. Furthermore, most horse carriers are experienced handlers of livestock and may be more gentle in handling your pigs than are carriers who typically haul pigs to the stockyard or slaughterhouses.

Reputable carriers will provide you with a list of references if you request it. In addition, ask at your local grain mill, feed store, livestock supply store or local truck stops for referrals, or post a Carrier Wanted notice there. You may find a local carrier who is making a delivery in the area where your pigs are located and can bring them back for you.

Carriers can also be found through the Internet or local trucking companies. Don't just shop for price; remember to check references. If at all possible, look at the vehicle and trailer that will be used to haul your pigs to be certain that they are in good and safe condition.

If you are purchasing your piglets from a long distance away, arrange for the breeder to ship them via airline in approved dog crates. This is a relatively simple process and a very good way to reduce the risk of exposing the animals to contaminants or illness. Furthermore, airline rates for long-distance shipping are often more economical than are overland shipping rates. However, airlines permit only piglets or very small pigs to

A simple hog carrier can be quickly assembled using two hog panels, bent to fit inside a pickup bed.

be shipped in regular pet carriers. Most airlines specify that the combined weight of your pig and its carrier cannot exceed 100 pounds. Each airline is governed by the USDA standards for animal safety and may have additional specific requirements for shipping as well as limited airport access. Start by contacting the airlines that are available at your local airport to see if they offer pet transport service. You may end up driving to the next closest large airport to receive your pigs.

Large pigs can be transported by airline in crates custom built for the pigs, but they are shipped through the cargo department, not at the passenger counter the way a pet is shipped. Most airlines have strict rules and requirements, including a background security check of the shipper and a minimum monthly charge. This would not be an economical way of getting only a few pigs to their new home.

When shipping pigs of any size, be prepared to pay for health certificates, any health tests required by your home state, the shipping crate, and the freight charge. Only use a known person, reputable carrier, or major airline for transport.

You can, of course, make the trip yourself to pick up your pigs. If you are transporting the pigs yourself, have proper bedding and water available for the trip. If you are traveling over several days, feed must also be brought along.

Piglets especially must be kept out of drafts. Small pigs can be transported in dog crates in the back of a pickup truck bed. Larger pigs can be transported in any livestock trailer. A pickup with a properly ventilated shell can also be used.

Welcoming Your Pigs

Have your facilities ready to receive the pigs upon arrival. Isolate your new pigs from any pigs already on the farm. Worm and vaccinate them right away if not done just prior to transport.

After a trip on a carrier or trailer, a pig will be thirsty. Provide water immediately to prevent dehydration.

Pigs will set about exploring their new environment right away. Rooting, sniffing, and touching are just some of the ways in which a pig familiarizes itself with its environment.

Provide water immediately. Allow the pigs to acclimate to their new surroundings for a short time before feeding them. This will allow the pigs to settle down and familiarize themselves with the area. When moved into a new place, pigs tend to jump around, spin, and trample through any dishes or obstacles they are not familiar with.

Feed your new pigs the same type of feed they have been eating. (Be sure to inquire about this from the seller. You may be able to purchase a few days' worth of feed from the seller to start your pigs off at home.) If you change the feed, begin mixing the old feed with new feed on day two, and gradually increase the proportion of new feed to help the pigs adjust to the diet change.

If you are raising purebred pigs, join the appropriate association for your breed. Get your pigs registered right away, so this important step does not get overlooked. Properly registered pigs are of more value and will bring higher prices if you sell them.

CHAPTER THREE

Pastures and Pigpens: Housing and Fencing Pigs

It's the big day, and you are bringing your pigs home. Having a well-prepared pig house and fences will save you many hours and headaches once your pigs arrive, so planning ahead is a must.

Hundreds of years ago, areas were fenced to keep the pigs out, not in. Pigs were left to roam the property, and the gardens or living areas were fenced to keep the pigs from rooting or destroying them. This may still be an appropriate method for some farms, but today most farmers will need to fence in or otherwise confine their pigs in some way to prevent them from trespassing on others' property. Pigs are crafty creatures, especially if they perceive food within reach, and they will go under, through, and over many barricades in search of a morsel. Careful site planning and materials selection will keep you and your pigs happy.

Selecting a Site

Avoid locating hog pens in any area where water pools or the ground becomes excessively muddy during rainy seasons. A slight slope will help drain away rainwater and urine and keep your pigs more comfortable. You, too, will appreciate a dry, slightly sloped site at chore time.

Herds of pigs weighing 200 pounds should be allotted at least 20 square feet per pig, or approximately ten pigs per acre. If you build an indoor-outdoor pen, design it so that 40 percent is indoors and 60 percent is outdoors; that is, each pig should be allotted at least 8 square feet indoors and 12 square feet outdoors. A sow with a litter needs at least thirty-five square feet, and a herd of sows should be limited to four sows

Providing plenty of room allows the pigs to roam and forage at will. Ample crop cover keeps the pigs busy and fed, possibly preventing some rooting.

per acre (if open grazing). Remember that this is an absolute minimum. Pigs grow quickly, and crowding seemingly happens overnight. With crowding, you invite illnesses, fighting, boredom, and unpleasant habits. Give your pigs as much room as possible.

Did You Know?

According to a report from Texas Tech University, each housed pig should be allowed the minimum amount of space as described below:

Sow (500 lbs), lactating, with litter	35 sq ft
Nursery pigs (7–60 lbs)	6 sq ft
Grower pigs (60–125 lbs)	10 sq ft
Finisher pigs (125–230 lbs)	14 sq ft
Mature adults	14–20 sq ft

Fencing

If you raise your pigs outdoors, fencing is a primary concern. Hog fences should be at least 32 inches high for smaller pigs and up to 60 inches high for larger pigs. Fencing can be constructed of many materials, including wood, hog or cattle panels, pipe, field fence, electric wire, and barbed wire. Each has advantages and disadvantages, including cost, ease of construction, and longevity.

Before you purchase fencing materials, prepare a map of the area you wish to fence, including all measurements, gate placement, and post locations. Carefully consider your gate placement and the way in which your gate will swing. This will make moving the hogs from pen to pen or out to pasture much easier. Your measurements automatically create your shopping list for supplies. Once you have your measurements, you can compare the costs of different types of fencing and the anticipated longevity of each. Constructing a fence that costs a bit more to build but will last many years longer is a sensible investment.

Wood Railings

Traditional pigpens or sties were made of wood rails, and many viable building plans are still available through the USDA Cooperative Farm Buildings Plan Exchange, at http://www.cerc.colostate.edu/Blueprints/Swine.htm.

If you keep your pigs on a concrete pad, then wood railings are quite suitable. Old concrete pads, such as those used for grain bins or previous barn sites, can be recycled effectively for pigpens. Wood used for pigpens should be at least 2 inches thick for strength and mounted on the inside of the posts. Lumber should be untreated, as pigs tend to chew their pen materials and may be poisoned by the wood treatment used. Some manufacturers now offer nontoxic treated lumber, as is used in children's swing sets. Check the label before purchasing.

A variety of fencing materials can be used to confine pigs to the selected area. Electrified netting, high tensile, and electric wire are all seen here. All are portable and easily maintained.

Pens can be quickly assembled using cattle or hog panels and T-posts. Use taller panels for larger pigs and pigs that climb, and use short panels for small or young pigs.

Wood railing fence is highly attractive when new but can quickly become an eyesore if the wood is chewed by the pigs. Wood typically does not last as long as does pipe or field fence, depending on the activity level of your pigs. If you have a ready source for wood railing, it may be the most cost-effective material for your pigpens. For areas that are wood poor (such as the plains states), wood pens may be more expensive to build.

Hog Panels

An existing concrete pad also lends itself well to being fenced with hog panels. Hog panels are welded-wire fence panels, typically 33 to 54 inches high and 16 feet long. These panels have graduated openings, with small openings at the bottom and larger ones at the top. This helps keep piglets in and adults from climbing the fence.

Hog or cattle panels are relatively inexpensive and sturdy. Farm supply and even home improvement stores should have panels available. A 16-foot by 16-foot pen, constructed of four panels, can be installed in less than one hour, at a cost of about $125, including metal T-posts.

Hogs raised on dirt can also be penned with hog panels. Panels should be attached to wood or metal posts sunk into the ground at least 3 feet. Level off the ground before putting up your fence to avoid any gaps at the bottom that could provide opportunity for escape. Panels should be set at ground level or below. This can be accomplished by digging a trench, setting the panel into the trench, and burying the bottom 6 inches of the panel. Alternatively, you can mound up and pack dirt about 1 foot high on the outside of the fence. If you

Concrete bulk feeders, as seen here, are a convenient place to feed larger animals. Heavy and sturdy, the concrete bunks last for many years.

Bulk feeders hold between 100 and 3000 pounds of ground feed. The flap doors allow the pigs to self feed while keeping out rain and debris. These feeders can be filled directly from the delivery truck.

bury your panels, purchase the taller ones (54 inches). Even very small piglets can get over shorter fences if they want to.

A disadvantage of hog panels is that they do not stand up to excessive climbing or standing on the horizontal wires. The welds will eventually break, creating a hazard for the pigs as well as an unattractive fence.

Pipe

Pipe is another effective fencing method. It will last for many years, and cannot be chewed. Pipe needs to be set on sturdy wood or pipe posts. Welding is a suitable way to adhere the pipe to the posts. In areas with widely fluctuating temperatures, mount horizontal pipes with floating brackets to allow the pipe to expand and contract within the bracket without breaking the welds. If you have pipe available, you can make a very stout pen using recycled pieces. Used pipe is available through metal salvage yards. Depending on the cost of steel, pipe may be the least expensive option for some farmers. Be sure to select pipe that is sufficiently large in diameter to withstand the pressure and weight of your mature pigs.

Field Fence

Field fence, or woven wire, is yet another fencing material that should be considered. It makes a very sturdy fence that will last for many years (some up to 50 years), but it also may be the most expensive type. A roll of field fence is 330 feet long and can fence an approximately

80-foot by 80-foot pen. This fence must be secured on posts set about 8 feet apart, with wooden cross braces set at the ends. Field fence comes in different gauges (weight of the metal wire used) and is priced accordingly. Field fence can be challenging to install, as it is very heavy and flexible until secured to the posts. If you are fencing a large area or a pasture for your pigs and feel certain that the fence will remain in place for many years, then field fence may be your most economical choice.

A pig can be trained to walk on a harness or leash. Clementine, a Mulefoot pig, is seen here taking a stroll around the Norfolk Virginia Zoo.

Electrified Fence

Pigs train well to a hot wire or electric fence. If you intend to pasture your pigs and move them to fresh pasture regularly, an electric fence is a very effective method of keeping your pigs contained. Two hot wires set at 6 inches and 18 inches off the ground can contain pigs of most sizes. An additional wire at 30 inches will keep in larger pigs. The wires should be energized by separate chargers for each wire. A perimeter fence is still recommended in anticipation of fence failure, power outage, or adventurous pigs. You should check your fence every time you feed or look in on the pigs to be certain that the charger and wire are fully functioning. Pigs can quickly discover a nonfunctioning wire and take advantage of the situation.

All pigs will enjoy some fresh air and sunshine. A happy Hampshire enjoys the warmth of the sun in an open-air paddock.

NetWire is an electrified mesh fence made of nylon and wire that comes in various lengths and heights. The wire is premounted on plastic posts, and it is lightweight and easy to move to new locations for rotating pastures. NetWire is also a good barrier to predators and can save your piglets from marauding foxes and coyotes. The cost of NetWire is comparable to that of field fence, but it is used primarily for temporary fencing situations. NetWire must be moved frequently to prevent grass and weeds from growing up through the netting, ruining the fence and preventing it from functioning.

A hot wire is also effective in keeping pigs from rooting under other types of fencing. A single wire set 6 to 8 inches off the ground, 1 foot in from the existing fence, can prevent your pigs from digging out. Keep in mind that pigs are very smart. Once they have been zapped by the wire, they will stay back at least a foot or two from the fence. This additional area should be considered when establishing pen size, as it will make the utilized area of the pen much smaller. When using hot wire, you should provide a gate that is not electrified. This will make moving your pigs much easier, as they will not willingly cross an area that was previously barred with a hot wire.

Barbed Wire

Barbed wire can be used to confine hogs to certain areas. However, the best use of

Advice from the Farm

Keeping Pigs Outdoors

Our experts give advice about raising pigs outdoors.

White Pigs

"White pigs need to be gradually introduced to outdoor living, as they will sunburn easily. A sow exposed to too much sun will sunburn. The sunburn causes the body to release prostaglandins, which will cause abortions."

—Robert Rassmussen,
Rassmussen Swine Farm

Reduces Parasite Load

"The most beneficial thing from raising pigs outdoors is that the fecal matter is spread out over a greater area. It reduces the manure and parasite load. Hogs need high-quality vegetation to get the most gain. Give them plenty of room."

—Josh Wendland,
Wendland Farms

Healthier Pigs

"Pigs raised outdoors are more healthy and it is more natural for them. Hogs need concentrate feed even if on pasture; they need some grain for growth."

—R. M. Holliday

Rotational Grazing

"Rotationally grazing pigs will give them the benefit of fresh air and sunshine, and will virtually eliminate odors. Grazed pigs are happier pigs because they can behave as they would in the wild, moving about or lounging around at will. Confining hogs to the same place over long periods of time fosters disease and reduces growth, productivity, and mental well-being of the hogs."

—Bret Kortie, Maveric
Heritage Ranch Co.

barbed wire for hog fencing is only as an adjunct to existing fence. For example, placing a strand of barbed wire 6 inches off the ground to the inside of an existing fence may prevent digging. A strand across the top of hog panels or wood fence may keep the pigs from climbing.

Rotational Grazing

A rotational grazing system is one in which the pigs are grazed in one area and then moved to another to allow the first pasture to regenerate. Rotational grazing of your paddocks will help diminish parasites, allow for fresh grazing and air, and improve your soil. Pigs are curious creatures, and a change of scenery also gives them something new to explore. Any large area can be arranged in consecutive paddocks to allow the pigs to move from one paddock to the next with ease.

If a pasture or paddock area becomes muddy (an exception being a mud hole described below), damp, or filled with feces, the pigs must be moved. These conditions are an open invitation to infection and disease in your pigs. Moreover, allowing your pasture to become overgrazed or overrooted will destroy the plant root system, and you will have to reseed it. Any area that is overused by pigs will be subject to these problems and should be avoided.

Pigs and Mud

In warm weather, pigs need access to clean mud to reduce their body temperature. You can help keep your pigs cool by creating mud holes for them or installing a concrete wallow or even a kiddie pool.

Tethering

Tethering is an alternative system that is halfway between grazing and penning. This is a common practice in the United Kingdom. The pig is tethered to a chain via a harness, similar to a dog-pulling harness, that goes around the pig's neck and chest, which is then anchored to the ground with a corkscrew rod. A swivel is attached to one or both ends of the chain to keep the chain from binding. Sows with litters are typically tethered while the piglets are left to roam. Piglets do not go far from their food source, so there's no need to worry. If you believe you have a secure area (where the pigs would be safe from predators), then tethering may be a good way to keep track of a few pigs. Once an area is grazed, the pig can be moved to another location. Sows with piglets under three days old should not be tethered, as the chain may be dragged over the babies and cause them injury.

A curious and mud-crusted piglet peers at the camera. Mud holes provide cooling and protection from biting insects.

Mud Holes

If pigs become overheated, they will make every attempt to create a mud hole by tipping their water dish or digging deep enough to reach cooler ground. Neither of these habits is beneficial to pastures or to paddocks. You must determine the spot where a mud hole will be situated. Give the pigs a helping hand by dumping water there for them. A sprinkler set up just above the area of your choice is also a good method. Be sure to place it over an area that can handle water, such as concrete with drainage or a mud hole you have selected where you don't mind water collecting. Running the sprinkler a couple of times a day will give the hogs a sufficient cool down.

Concrete Wallows and Kiddie Pools

A concrete wallow is a very shallow swimming pool intended for water, which may be more sanitary and manageable than a mud hole. Make sure this concrete mud hole has sloping sides to allow the pigs to get into and out of it easily. The depth should be no more than one foot for adult pigs and three to six inches for smaller pigs to prevent drowning. You can equip the concrete mud hole with a drain for easy cleaning: prior to pouring the concrete, install a standard tub drain connected to PVC pipe extending beyond and draining away from the mud hole. Be certain that your drain releases the dirty water onto a slope away from the mud hole and onto gravel or well-drained soil to prevent creating a mudslide. Gravel around the mud hole will help maintain the area as well as provide drainage.

Give some thought to placement of the wallow for maximum utilization. If you would like to use the mud hole for more than one pen of pigs, you need to place it where the pigs can be moved easily into and out of the area. Concrete mud holes are not practical for every

A concrete wallow is a wonderful place for a pig to cool off. Keep the water level shallow enough that pigs can lie down in the water without drowning. Just a few inches, as seen here, is plenty for small pigs.

Hose 'Em Down

A simple garden hose can be used to cool your pigs as well. Once they figure out that you are cooling them and not chasing them with the water, they will come running to get their shower. Many pigs will lie down as soon as the water hits them and turn their faces directly into the stream.

pen, as they can be expensive to build and will likely allow access to only a few groups of pigs at a time.

Plastic kiddie pools are great fun for pigs as well. If your pigs are not terribly destructive, a child's pool is a great place to cool themselves and lounge around in. As with the wallow, do not fill the pool more than 1 foot deep. A kiddie pool is not recommended in areas with small piglets, as the straight, rigid sides will prevent a piglet from getting out.

Barns, Huts, and Pens

You must provide adequate shelter for your pigs. A shady spot in summer and a shelter from the wind and the snow in winter are the minimum requirements. Shelter need not be elaborate or costly, however. Tree stands provide great shelter for hogs during all seasons. Simple A-frame structures, mini–hog huts (Quonsets), or an open door to the barn are sufficient for shelters.

A-Frames

A-frames, or arks, are simple structures, usually made of wood, with steeply sloped sides that allow for drainage and snow shed. Another advantage of the A-frame's shape is that the slope provides cozy interior areas where piglets can lounge without being lain on by the sow. A-frames can have elaborate door structures or just an opening no more than half the size of one end. A burlap bag or similar cloth can be stapled across the top of the opening to cut down on drafts. Wooden floors are typical in A-frames, although a stout base can be left open, allowing for drainage under the hut. Wintering in A-frames is best done with a floored unit.

Quonset Huts

Quonset huts are corrugated steel structures shaped like an inverted U. One end is walled in, with an adjustable vent at the top. The other end is left open or partially walled to make a doorway. Quonset huts are fairly lightweight and need to be secured to the ground with rebar rods. They are easy to move.

Mini Quonset huts provide ample space and protection for the nursing sow and her piglets. Piglets will soon venture out of the hut but will return for sleep, warmth, and safety.

Quonsets can also be outfitted with many extras, such as a guard rail around the bottom to provide a safe place for piglets, where mom can't get in and accidentally lie on them.

A-frames and Quonsets can be fitted with pens or small fenced areas to allow the pigs some exercise while containing them for various reasons. Either could be mounted on skids or wheels and moved about to allow access to fresh pasture as needed.

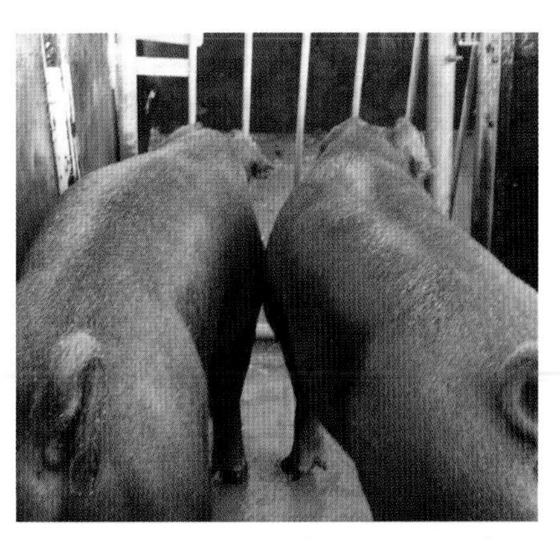

Secure pens keep pigs where you want them to be and allow easy access by the herdsman. Metal rungs can withstand years of chewing and rubbing without any damage to the metal.

Existing Structures: Barns and Pens

Nearly any existing barn can be retrofitted to accommodate pigs. Old stables, lambing jugs, cattle pens, chicken barns, and the like can be modified to provide adequate or even optimal hog facilities. Horse stables usually require little more than adding a few extra boards along the bottom of the pen. Lower ceilings will provide more warmth in winter. Pigs do not like to traverse steps, so a ramp or opening directly into their pen is advisable.

Calf huts make excellent housing for pastured pigs. Growers, farrowing sows, and the breeding herd can all be housed inexpensively using these structures.

Doors should be a consideration when building pens for pigs. A door can serve as a traffic control or stop when moving pigs around. In addition, the way a door opens can make cleaning chores easy or exasperating. Consider a swinging door that has removable hinge pins on each side, allowing the door to swing left or right. Secure locks, gate hinges, or hooks prevent the pigs from opening the gate by either lifting, climbing, or rubbing on it.

Building Your Own Structure

If you are considering raising large numbers of hogs indoors, then building dedicated, customized pig housing is advised. Remember that pigs do best when they get fresh air and sunshine, so a barn with an outdoor exercise area is highly recommended. If you are designing a larger setup, evaluate several design plans that compare function, ease of feeding and moving the pigs, comfort of the pigs and stockperson, and cost. Many small-scale plans are available through county extension offices, USDA bulletins, and private companies that manufacture and build hog units. A variety of plans can also be found through old livestock books and on the Internet.

Site location for such a unit is critical for several reasons. You will likely need to move pigs in and out, and you will need access to the building with trailers, trucks, or crates. If your feed is delivered, the delivery truck will need to be able to get close enough to the storage area to deliver feed without running into anything or getting stuck in the mud. A well-drained site situated far enough away from other obstacles will make the building highly functional.

Bedding and Ventilation

It can't be said enough that pigs need dry bedding at all times, whether they are on pasture or in houses. Dry bedding makes the pigs more comfortable and keeps them warmer in winter months. Bedding also absorbs moisture

Deep bedding is very comfortable for pigs, as it will keep them warm and dry and provide a cushy place to lie. The bedding composts from underneath, creating more heat for the pigs.

and odors, making the whole environment more pleasant. If you are using portable huts, the bedding can be placed in the hut and then spread out over the pasture when the pigs are moved to a new area. Bedding dispersed over an area will provide compost and break down much more quickly than if left piled up.

A thick layer (about 4 to 6 inches) of wood chips or sawdust covered with a 2-inch layer of straw makes fine bedding. Other materials for bedding can include old hay (not moldy), wheat middlings, rice hulls, cedar chips, cornstalks, and commercial pellet bedding. Pigs like to nibble and chew on bedding, so a nontoxic material is a must. A decent layer of bedding will keep the pigs dry and free from odors and will also make stall cleanup easier.

Ventilation is likewise a vitally important issue in housing pigs. Proper ventilation moves air without causing a draft on the pigs and keeps moisture and ammonia odors from building up. If you choose to house your pigs, opening a door a few hours per day may be enough to solve your ventilation issues. Windows can provide light and ventilation if installed a foot or so above the height of the pigs. Fans installed in the peak of the ceiling will also help circulate air without causing drafts directly on the pigs.

CHAPTER FOUR

Feeding Your Pigs

A proper diet and adequate water are the main source of care you need to give to your pigs. Feed and water should be given in separate dishes, and the pigs should have access to fresh water at all times.

Water

Water is the most important staple in a healthy pig's diet. Under moderate weather conditions, a pig will consume 1/4 to 1/2 gallon of water for every 1 pound of dry feed. During hot weather and while lactating, a pig's water needs will increase to 1/2 to 1 gallon of water per 1 pound of feed. The chart on page 68 estimates typical water consumption at various stages in the pig's life.

There are several methods of watering pigs using more or less complicated equipment, including water dishes, troughs, nipple drinkers, hog watering tanks, and automatic heated waterers. The type of waterer or dish you choose will have a large impact on your chores and overall satisfaction with raising pigs.

Water Dishes, Troughs, and Nipple Drinkers

The water dish is perhaps the simplest watering method. Water dishes are relatively inexpensive. Made of rubber, metal, concrete, or durable plastic, they can be moved easily or tied to a post or fence. Remember, however, that pigs love to turn things over with their noses, and a water dish is a favorite target. If you don't mind filling the water dish several times per day, this may be the right choice for you.

A trough is a more efficient choice if you are watering several pigs. A trough can be made by splitting an old hot water tank in half lengthways and either welding legs

Water Needs

Life Stage of Pig	Estimated Water Requirements (gal/head/day)
Nursery pig	one
Growing pig	three
Finishing pig	four
Gestating sow	six
Sow and litter	eight
Adult boar	eight

to it or anchoring it to some cinder blocks or 4-inch by 4-inch blocks of wood. Readymade troughs are available from several manufacturers of livestock equipment. These troughs can be used as is and moved easily or bolted to concrete or otherwise permanently anchored. Other possibilities for troughs are hollowed-out hand-hewn logs and glazed half pipes (as used in sewer drain lines).

Metal tanks with an automatic drinking hole in the side are very efficient during summer. Access to small amounts of water through the hole keeps the remaining water free from dirt and debris.

A nipple drinker is another watering option. Piglet drinkers come in one-gallon tanks with a nipple on the side. The nipple is a metal valve that releases water when pressed in with the pig's nose. This is a fairly sanitary way of watering small pigs, as the animals are not able to step in or play in the water. You must be certain to anchor it securely to the side of the pen or a post, as the constant pushing on the nipple will wiggle the unit loose. As the piglets grow, the unit can be progressively raised (hung higher) to always keep the nipple at the same height as the piglet's back.

Nipple drinkers can also be installed on readymade pipe fittings. These L-shaped pipes are prethreaded to accept the drinker. They have an adjustable holder so they can be used with any size pig. You then hook the top end to a permanent water line or pressurized garden hose for water. These are really nice in the summertime or for heated barns but are not suited to outdoor facilities or non-heated barns in the winter, as the lines will freeze and break.

Watering Tanks and Automated Waterers

Hog watering tanks are sold in a range of sizes by several manufacturers. They typically hold between 35 and 250 gallons of water, and come equipped with a hog drinker, which is a trough cut into the side. These automatically refill via a float valve. Tanks can be equipped with heaters for winter months. Most watering tanks are challenging to keep clean. The drinker portion is inset into the side of the tank, making it nearly impossible to clean out without emptying the whole tank and turning it upside down. In certain environments, such as on concrete or inside barns that are not bedded, watering tanks are ideal. In dirt situations, the drinker portions tend to fill up with sediment, as the pigs basically wash their faces in the drink well while getting to the water.

An advantage to watering tanks is evident when medicating pigs or when vitamins or supplements need to be added to the water. The tanks hold a set number of gallons, making it easy to mix the proper amount of supplement.

Automatic heated waterers are the most expensive but least labor intensive method for watering pigs. These are permanently installed on a small con-

Permanently affixed, stainless steel hog drinkers are virtually maintenance- and trouble-free. The pigs help themselves by lifting the flap doors. Flaps help to keep debris out of the waterers.

Advice from the Farm

You and Your Pigs

Suggestions from our experts regarding your relationship with your pigs.

Develop Pig Sense

"A person needs to develop pig sense. By getting to know your pigs and their behavior, you will recognize when a pig is telling you that something is wrong. A pig lying off by itself or not eating is a sign of a problem."

—Al Hoefling, Hoefling Family Farms

Know Your Pigs

"All pigs have individual personalities. Get to know them as individuals, and working with them will be much, much easier. Knowing who you can pet, move, or walk without restraints will help you to work your pigs with less stress—on you and the pigs."

—Bret Kortie, Maveric Heritage Ranch Co.

Check Them Often

"By hand feeding and watering your pigs twice per day, you give yourself the opportunity to look over your pigs and pens and recognize and correct any problems right away. This will save many pigs as well as equipment and money."

—Robert Rassmussen, Rassmussen Swine Farm

Off Their Feed

"Check out any pig that doesn't come up to eat. Any animal standing off by itself is not a good sign. Look for the source if any pigs are coughing, sneezing, or such. Dusty pens should be watered down to keep the dust down. If they can be treated and remain with the herd, that is best, as no animal wants to be isolated."

–Josh Wendland, Wendland Farms

Bulk feeders with protective doors or flaps are an efficient method of feeding large numbers of pigs. Pigs can have access to feed at all times. Feeders can be filled with enough feed for many days, saving labor and time for the herdsman.

crete pad, plumbed, and electrified. The water flows automatically via a float valve. Most come with a flip lid that the pig lifts up while drinking. This lid prevents most contamination from feces or bedding. If your pigs are not familiar with a flip lid, you can tie the lid open for the first day or two so they find the water before learning to use the lid. Cleanup is fairly simple, as the waterers are equipped with a drain plug. By removing the plug, water flows through the drinker and out the bottom, flushing the drinker well clean.

Feed

Many homestead hog raisers support feeding pigs on the ground. Scattering the feed around the pen encourages the pigs' natural behavior, as they move about looking for morsels just as they would in the wild. If your pens are kept relatively clean, and you feed in the feeding area—not the bed or toilet areas—then this is a very practical method. Only large grains, cubes, or pellets work sufficiently, though. If you are feeding finely ground feed, you will need to provide troughs to avoid high waste. Finely ground feed is lost among the dirt and gravel and may be trampled into the same.

Feed Dishes and Troughs

Feed dishes have the same drawbacks water troughs do—pigs love to overturn them, so heavier is always better. Metal troughs are durable, are relatively easy to clean, and can be secured to a fence or wall. Troughs must be divided into sections if used for more than one pig, however, to prevent the dominant pig from standing in the trough or pushing the other pigs out.

Grower pigs should have access to feed at all times to encourage consumption and growth. Regulating what each pig eats is challenging with this system. By regulating the flow of feed, you can minimize excessive eating.

Wall feeders and troughs are suitable to small-scale production. They are easy to fill and clean and can provide ample room for up to six pigs at one time.

Wall Feeders and Bin Feeders

For at least until six weeks old, piglets should have wall feeders within a creep—a portion of the pen that is walled off from the sow. The creep typically houses the starter ration and water and may include an extra heat source for the piglets. This allows the piglets to feed, water, and rest without the threat of being accidentally crushed by the sow and without having to compete with the sow for food. Piglet feed, or starter feed, is typically finely ground or compressed in small pellets, and it needs to be placed in a feeder at a height suitable for small pigs. Piglets can also be fed from shallow pans, but much feed will be wasted by trampling and spilling. (See the chart on page 83 for specific creep-feed requirements for pigs.)

Wall feeders are also made for larger pigs, and they can be a clean and easy way to take care of small numbers of hogs. Wall-mounted hay feeders designed for horses are also functional for hogs. They are sturdy, can be mounted at any height, and can be filled easily with hay, clover, or alfalfa.

Bulk bin feeders are another option. This apparatus costs more than other feed equipment but offers significant labor savings. Also called a self-feeder, a bulk bin can hold from 50 pounds to several tons of feed at one time. The trough portion is a series of flap doors the pigs can lift when they want to eat. An apparatus within the bulk bin can be adjusted to release the amount of feed needed, so that only the proper amount is accessible to the pigs at a time. The pigs use their snouts to work the feed out from under a bar inside the trough. This style of feeder is especially efficient when raising market hogs, which should eat a large amount frequently: you can adjust the feeder so a larger portion of feed is readily available in the trough at all times. Adult pigs may overeat on this type of feeder, so adjusting the ration is important.

What Do I Feed My Pigs?

This brings us to the big question: what do I feed my pigs? This topic seems to cause first-time pig owners more anxiety than any other, but the simple answer is

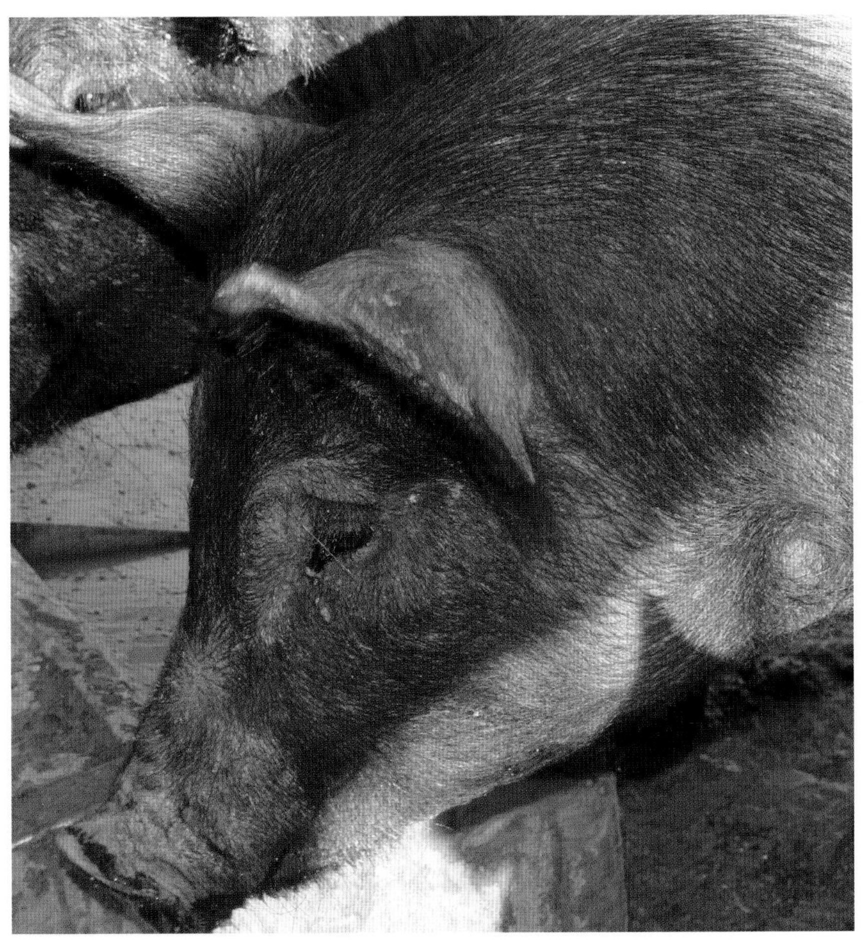

Allowing grower pigs access to feed at all times eliminates some of the competition and fighting over food, reducing stress in the pens. Once the pig learns that he can have food any time he wants it, he will relax and enjoy.

that pigs will eat and do well on a variety of feedstuff. Remember that pigs in the wild have survived for thousands of years on whatever food they could find. Once domesticated, pigs were fed whatever the farmer could provide. Today, an extra few rows in the garden, lawn clippings, kitchen scraps, surplus vegetables, milk, and crop residues are all viable and nutritious swine feeds.

That said, there is a large body of work devoted to the subject of pig feed. Dozens of books have been written on the subject, and major universities have formal research programs dedicated to analyzing and developing hog diets in an effort to improve pork production. Kansas State University and the University of Minnesota both have publications available through their extension offices on the formulation of feed for various stages of a pig's life. (See the Resources section in this book.)

This highly technical and advanced knowledge of hog feed formulation certainly contributes to better animal hus-

Rotten and Moldy

Rotten food and moldy grain can make pigs sick. Mold on grains can contain aflatoxins, poisonous substances that cause neurological impairment, abortions, internal bleeding, and death. Although pigs are not as susceptible as humans and other mammals to contaminants in food, it is far better to recognize and eliminate contaminants than to have to treat a sick animal or lose one later. Piglets are especially susceptible to toxins in mold, and piglets can suffer from bloat, liver damage, and death. Often, contaminated feed is not recognized until after it has caused sickness or death.

Molds and toxins in feed may not be obvious to the naked eye. Mycotoxins can be seen under the presence of a black light—they will look fluorescent green. Barring black lights or sick pigs, a lab is a sure way to identify contaminants in the feed. Most universities have laboratories that can test feed for mycotoxins and a variety of other food contaminants. A sample of the feed can be sent to the lab for testing. If you have symptoms in the pigs, you should notify the lab of all of these symptoms to enable them to test for specific toxins.

Purchasing your feed from reputable granaries or feed mills that routinely test their feed for contaminants is your best option. Do not knowingly buy grain from sources that store the feed outdoors, in leaking bins, or in damp or wet areas. Feed that is obviously moldy, has any peculiar odor, or looks dirty should be strictly avoided.

A pig on pasture is a happy pig indeed! Seen grazing, this weanling pig is enjoying the variety of pasture and the adventure of exploring new territory.

bandry practices, but you do not need to be an expert in the subject or seek out specialized suppliers to find the right feed for your pigs. An advantage of raising pigs on a homestead or hobby farm is that you will be able to produce or supply a large amount of the feed necessary to raise healthy pigs. Developing your own feed ration is truly not a complicated process. Armed with knowledge of your pigs' basic dietary needs (described later), you will find that feeding your pigs can be as simple as you want it to be. A proper ration will have the right balance of protein, fiber, vitamins, minerals, and carbohydrates.

Seasons and stages of life alter the nutritional needs of your pigs. When on pasture, pigs may need supplemental protein or nutrients. Excessively cold weather may increase your pig's need for food, as it is using calories to maintain body temperature. Extremely hot weather can often decrease the feed intake of pigs, as they simply do not feel like eating if they are too warm.

Pasture Raised

If you are able to pasture your hogs, you will be satisfied with the growth and health of your pigs. Pasture raising adds many benefits to the pig's life as well as to

Gradual Transition

It is best to turn your pigs onto lush pastures for a few hours at a time to start. By gradually adjusting the amount of time the pigs are allowed to graze each day, you will also allow their digestion to adjust to the increase in water and fiber in the diet. A gradual introduction to succulent feed will prevent diarrhea in most hogs.

the quality of the meat. By grazing, pigs are able to exercise and perform many natural behaviors such as rooting, wallowing, and exploring. Your meat quality will be improved by the addition of healthy grasses and plants, healthy omega fatty acids from these plants, vitamin D from the sunshine, and fresh air. Pastured pigs have little odor, as the wastes are not concentrated in any one area.

Well-managed rotational pastures can provide up to two-thirds of a hog's food ration. (Supplemental grain or milk is usually required for the remainder of the ration to assure proper nutrition and normal growth.) Popular pasture mixes include orchard grass, alfalfa, and clover. Winter vetch, oats, and peas are also a desirable pasture mix for pigs.

Good Pasture Land

Good pasture land for pigs includes an area with a wide variety of plant sources. Legumes such as alfalfa, peas, and clover provide an ample amount of protein in the diet. These plants are also

A variety of plant materials provide many nutrients for your pigs. Seed heads, weeds, and roots all have differing nutrients that support healthy growth and maintenance of your pig herd.

Poisonous Plants in the Pasture

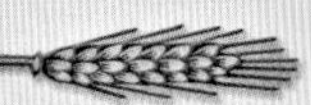

Before turning your pigs loose on pasture, inspect the area for weeds that can be poisonous to your pigs. Some common poisonous plants are pigweed, Jimson weed, two-leaf cockleburs, lambs-quarters, spotted water hemlock, pokeberry, black-eyed Susans, and nightshade. Do not assume that your pig will not eat anything poisonous because it tastes bad. Pigs are not highly discriminatory when it comes to feeding, so it is up to you to safeguard your property. The ASPCA National Animal Poison Control Center (see the Resources for contact information) maintains a current list of plants that may be poisonous to your pigs. If you cannot identify the plants on your property, contact your local county extension office, USDA office, or Natural Resources Conservation office, or ask the plant science department of one of your state's colleges for assistance. Check the phone book or Internet for phone numbers and addresses of the various agencies. Most states have catalogues of plants native to the state available to residents, including drawings or photographs, plant descriptions, uses, toxicity, or palatability.

highly palatable to pigs, so the pigs will eat them voraciously.

Grass varieties provide nutrients, fiber, and variety in the hog diet, but various grass species contain differing nutrients. Typically, stemmy and thin grasses have less nutritive value than do lush, thick grasses, although this is not always the case. A purely grass-based pasture will not provide all the nutrients your pigs need, and they will require supplemental feed. Grass also contains high amounts of water and fiber and needs to be balanced with other feeds.

Nongrass forbs—flowering plants with nonwoody stems, such as wild radish—are also highly nutritious and palatable to hogs.

Aside from the nutrients your pigs derive from the pasture plants themselves, your pigs gain added nutrients from the soil. Soils vary greatly across the United States, but most contain trace minerals such as iron, copper, and zinc, as well as salts and vitamins. An added bonus is that the pigs will also acquire added protein as they root up bugs, worms, grubs, and small rodents.

Rotational Grazing

If you pasture your hogs, grazing rotation is essential. Rotating your pigs to various pastures will allow areas to regrow, will give sufficient time to allow the composting of manure, and will prevent your pigs from completely rooting

These piglets are not only benefiting from the varied plant life but are also gleaning grains and alfalfa left behind by their pasture mate.

up and destroying the area. Rotating also helps control internal parasites.

A pasture full of thistles can be reclaimed for better crops by running hogs on it until it is rooted up. Hogs will dig down along the side of the tap root in a thistle, munch off the root, and leave the top to die. This is a great healthy alternative to herbicides and other chemical treatments. Thistles are also a highly nutritious and healing plant and can increase your pigs' vitality.

Pastures and grass paddocks can be set up in rows with gates between them to make rotating an easy task. Another clever method is to create a circular set of paddocks divided into pie-shaped wedges, with the water and feed at the center. This makes moving the pigs a simple task. Chores are more manageable because the water and feed are at the center and do not have to be moved when the pigs are moved.

Once a pasture has been grazed by your pigs, you can seed the area with other varieties of vegetation such as clovers and small grains. By allowing each paddock to rest for at least three weeks between grazing, each will have the opportunity to grow additional healthy vegetation.

Grazing Also Benefits the Land

By allowing your pigs to graze openly, you will be adding benefit to your land.

Pigs will be naturally fertilizing the area on which they graze, returning valuable nutrients to the soil. By selectively grazing certain areas, you will be able to eliminate some noxious plant species. Moving your pigs as needed will allow the more favorable species to recover and regrow.

Pasturing: Alternative Feeds

In some areas, pasturing is practical only during certain seasons of the year or when there is ample rainfall. When pasturing is not feasible, you must provide alternative feeds. The first step in finding alternatives is to look to the feedstuff available in your area. Alfalfa, clover, and grass hay are all beneficial feed sources for pigs. If your neighbors are growing wheat, barley, triticale, milo, oats, or corn, you should start with a ration from these grains. Buying local is always the best method; it will save you money and time, and you will be supporting some of your neighboring farms.

Many antique texts relating to swine husbandry refer to the Trio as the best feed for hogs. The Trio consists of milk, alfalfa (fresh or hay), and a grain mixture. If you are able to feed your penned pigs with this ration, you will have healthy, well-nourished pigs.

Commercial Rations

When pasturing is not possible and penning is necessary, the most convenient way to feed your pigs is with a commercially formulated balanced ration. These rations are blended for the age and weight of your pigs. Most feed stores carry only rations that contain antibiotics or other unnatural ingredients, such as choice white grease (which is pig lard), chicken feathers, or other animal waste. Cost and questionable ingredients may eliminate these as feed types for many homesteaders who seek to raise pork that tastes different from commercially raised animals.

Organic feeds are becoming more readily available, as the demand for these feeds grows. Organic feeds are certified to contain only organic ingredients and are not permitted to have any animal by-products, such as bone or fish meal. Measures are also being taken to eliminate any genetically modified grains from rations certified as organic. Organic feeds are free from antibiotics, pesticides, organo-phosphates, and chemical fertilizers.

A comparable ration can be mixed by your feed mill, whereby you can determine the ingredients and amounts yourself and avoid unwanted ingredients. Moreover, custom mixes are usually a lot less expensive than premixed rations are. You need to provide a secure storage method for the feed, such as lidded barrels or large trash cans, a bulk grain bin, or other rodent-proof container. Buying custom mixes by the ton will also save you money. If you lack storage containers or cannot use a ton of feed within a two- to three-month period, you could buy smaller amounts and have the mill bag the ration for you. Bags must be stored where they will not get damp or wet and cannot be chewed into by rodents. Bags typically weigh 50 pounds each.

Grains

Grains are the most common hog feed in the United States, and at this time the most easily acquired source of feed. Most commercial producers and many small farms feed grains as the only feed source to pigs. Even pastured hogs can benefit from a supplemental grain ration when fed at 40 percent of the total feed ration. Grains provide a variety of nutrients, fiber, carbohydrates, and bulk in the diet.

Corn

Corn is the most common ingredient found in hog feed in the United States. Of traditionally grown grains, corn is the second richest in fattening food nutrients (wheat is the first). Corn is an excellent energy feed and is an especially good finishing feed, as it contains a high amount of digestible carbohydrates and is low in fiber. Corn is also used in the diet to add firmness to the fat on a pig.

In spite of all the benefits and virtues of corn, it will not produce healthy hogs all on its own. Corn contains between 7 and 9 percent protein on average but is deficient in nearly all amino acids required for growing, gestating, and lactating pigs. Corn is also deficient in calcium, minerals, and vitamins. Pigs fed on corn alone will eventually die from malnutrition and related diseases. Therefore, corn must be supplemented with vitamins, minerals, and a protein (amino acid) source. When properly supplemented, corn is an excellent feed for all ages of swine.

Wheat

Wheat is considered to be equivalent to corn as a feed source. Wheat is slightly superior to corn in quality and quantity of protein, at 11–14 percent, and contains more lysine (an essential nutrient for growing and gestating pigs). Wheat may appear to be more expensive than corn, but consider that corn at $3 per bushel weighs 56 pounds. Compare this to wheat at $3.21 bushel, which weighs 60 pounds. Wheat can replace corn one to one in feed rations, and it will decrease the amount of soybean meal or other protein required to meet the desired protein level. Wheat should be rolled or coarsely ground, as finely ground wheat may form a pasty mass in the mouth and become less palatable than corn to some pigs.

Oats

Oats are not considered a high-energy feed. Whole oats have about 30 percent fibrous material, mainly from the hull. Although fiber adds bulk to the diet, it is of no nutritive value to hogs. Oats fed to fattening and grower hogs should not constitute more than one-third of the total ration. On the other hand, oats are an ideal ration when fed to sows in late gestation and early farrowing, as the added bulk helps keep the sow regular and feeling full so she does not overeat.

Hulled oats are an excellent feed for any size pig, although this extra processing step does increase the cost of the feed. This feed may be practical for getting weanling pigs off to a good start or when fitting a pig for show.

Barley

Barley can replace corn in a ration up to about 75 percent. Barley is less palatable to pigs and should therefore be mixed with the protein source and fed as a full ration to ensure that it is ingested. Comparable to corn in protein, barley also lacks the proper amino acid balance. Like corn, barley must be mixed with the proper supplements if it is to supply adequate nutrition. Scabby barley (barley affected by the *Fusarium* fungus) should never be fed to lactating sows or small pigs. Scabby barley possesses a vomitoxin, which when ingested causes vomiting and nausea. Additionally, *Fusarium* fungus causes liver, intestinal, and reproductive disorders in pigs.

Rye

Rye is considerably less palatable to swine and should not make up more than 20 percent of the total feed ration. Rye should be ground and mixed with more palatable feeds. Rye is often infested with the fungus ergot. Ergot causes dementia and abortion and reduces feed consumption and growth rates. Infested rye can be identified by the black kernels on the seed head or

When feeding confined pigs, commercial rations such as the pellets shown here provide all the protein, fiber, carbohydrates, and nutrients a pig needs. Feed should be matched to the growth stage of the pig for best results.

mixed within the rye seeds. If you are unsure about whether rye has ergot contamination, a laboratory can positively identify the toxin. Ergot infested grain should never be fed to pigs.

Sorghum

Sorghum (milo) has about 92–95 percent of the feed value of corn. Milo should be ground and mixed into the ration since the small, hard kernels may not be chewed sufficiently. Milo has nearly the same benefits and deficiencies as corn and should be mixed with the appropriate supplements.

Milk Products

Skim milk, buttermilk, dried milk, and other surplus milk products contain nearly all the nutrients that grains lack. Milk products are also highly palatable, easily digested, and of great value to recently weaned and growing pigs. Homesteads and hobby farms producing their own milk will have a ready consumer in the pig. A gallon of whole milk weighs 8.6 pounds, and skim milk weighs about 8 pounds. Feeding 3.5 gallons of skim milk is roughly equal to 1/2 bushel of corn, or 28 pounds. Adult pigs also benefit from the nutrients in milk and can be fed as much as 1.5 gallons per pig per day, in conjunction with a grain ration and fiber source.

Whey is the by-product of cheese production, the cloudy liquid that remains after the milk solids are rendered from the whole milk. Whey contains a high percentage of water and little protein. It is of little value in feeding growing or market pigs. It is best used for adult hogs as a supplement to grain and water. If you are lucky enough to be located near a dairy or cheese processor, whey is an inexpensive addition to a balanced ration and will greatly enhance the palatability of the grain ration.

Proteins

Grains contain too little protein and are deficient in too many vitamins and minerals to be sufficient as a complete feed. Proteins are the most significant category of nutrients in swine rations. Proteins are broken down into various amino acids during the digestion process. These amino acids assist the pig during growth, reproduction, lactation, maintenance, and healing. Amino acids cannot be manufactured by the body, so they must be provided in the form of properly mixed feed.

Overfeeding protein is a waste on all counts. Protein-rich feeds suitable for pigs are generally more expensive than are grains (per pound), so a combination of grain and protein puts less strain on the pocketbook. Typically, excess protein is excreted or adds weight to the pig's liver and pancreas without increasing muscle or back fat. Excreted protein has a very high concentration of nitrogen, a substance known to contaminate ground water. Finally, pigs are inclined to overeat protein supplements if fed free choice. Mixing the proper amount of protein into the grain ration is highly recommended.

Protein Charts

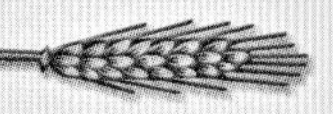

Average Amount of Digestible Protein in Various Swine Feeds

Feed	Percentage of Digestible Protein	Feed	Percentage of Digestible Protein
Alfalfa	16–17	Meat Scraps	50
Barley	9	Millet	8
Buckwheat	9	Molasses	>1
Buttermilk	3	Oats	9–10
Corn	7–9	Rye	10
Cottonseed Meal	38	Skim Milk, Dried	33
Cow's Milk	4	Sorghum	5
Cowpeas	19	Soybean Meal	38
Fish Meal	48	Tankage	56
Linseed Meal	30	Wheat	11–14
Meat and Bone Scraps	47	Whey	>1

Protein Ration and Gain

Category/Age of Pig	Recommended Protein Content (Percent)	Avg. Daily Gain (lbs)	Avg. Daily Feed Intake (lbs)
Grower/Finisher Pigs			
Suckling pigs, 5–30 lbs	22	0.7	0.5
Nursery pigs, 10–20 lbs	22	0.3	>0.5
Starter pigs (weaned), 20–40 lbs	18–20	0.6–0.9	1.1
Grower, 40–125 lbs	16	1.5	3.5–4.25
Finisher, 100–240 lbs	12–14	1.8–1.95	7
Gestating Pigs			
Gilts	14	0.6	3.5–6
Sows	12–14	0.35–0.45	3–5.5
Lactating Sows	16	0	12
Boars			
Under 1 Year	14	1.9	2–6
Breeding Boars	16	0	6

It is easy to determine the appropriate amount of protein to add to a grain ration. Determine the protein level you want to feed from the Protein Ration and Gain Chart. For this example, we will use corn as our base and soybean meal as our protein supplement. Let's say you want to feed 12 percent protein for a sow ration. Now, determine your base feed (corn) and its average protein level (9 percent) from the chart titled Average Amount of Digestible Protein in Various Swine Feeds. To find out how much corn and how much soybean meal you need to make a ration that is roughly 12 percent protein, follow these calculations:

Percent protein in chosen supplement (soybean meal)	38 percent
Protein level desired	12 percent
Subtract desired from chosen base required	Subtotal: 26 parts
Protein level desired	12 percent
Percent protein in base (corn)	9 percent
Subtract base from desired protein supplement	Subtotal: 3 parts
Add two totals together	Total: 29 parts total feed

Now, to determine how much grain and how much protein to use per every 100 pounds, take the amount of your base feed (corn), divide it by the total amount of feed, and multiply that by 100. In this example, 26 divided by 29 equals 0.9, multiplied by 100 equals 90. Therefore, 90 pounds of corn mixed with 10 pounds of soybean meal will contain approximately 12 percent protein.

Beans and Seed Meals

Soybean meal is probably the most economical source of quality protein for hog rations. It is highly palatable to pigs but should be mixed with the grain ration to prevent pigs from overeating the soybean and neglecting the grains. Soybeans are fairly well balanced in amino acids but deficient in minerals and vitamins. A vitamin-mineral supplement is an absolute must. A combination of soybean meal and alfalfa meal mixed 3:1 will also balance the ration.

Raw soybeans contain high quantities of trypsin inhibitors, which block normal protein digestion in pigs. Therefore, raw soybeans should not be fed to baby or weanling pigs. Raw soybeans and peanuts should not be fed to grower or finisher hogs (market hogs), as they create an undesirable carcass with soft meat and lard and fat of poor quality.

Linseed meal is well liked by pigs. It has a slight laxative effect and can be used efficiently in sow herds. Although it contains high amounts of calcium and adequate B vitamins, linseed meal is highly deficient in lysine. Supplementation is necessary. Linseed meal should not make up more than 20 percent of the total protein supplement.

Cottonseed meal contains a toxic substance called gossypol. It is also not very palatable to pigs and is not recommended.

Tankage, Meat and Bone Meal, Fish Meal

Other products commonly added to pig feed as protein or fat supplements are tankage, meat and bone meal, and fish meal. Animal proteins are typically more expensive than plant-based proteins are, although not always. Tankage is a product derived from rendering animal carcasses. The product is then separated into protein and fat, respectively. Typical protein levels in tankage are about 60 percent, so it can be added to rations in a very low amount to obtain the protein level desired. It is generally liquid in form.

Meat and bone meal and fish meal are derived primarily from the by-products of butchering and processing. All are typically ground and dried and added to feeds in a meal form. Fish meal has been known to add a peculiar (fishy or weedy) taste to meat if used in large quantities.

Not only are weeds delicious and nutritious, but they also make a pretty good place to take a nap.

Tankage and meat and fish meals are deficient in tryptophan and should not be used as a single source of protein. They are good sources of calcium and phosphorous, however. Although not as palatable as soybean meal to pigs, these meals can provide up to half of the protein supplements needed in a ration.

Carefully check the source of tankage if you wish to use it in your ration. Tankage may be the product of euthanized or sick animals or roadside carcasses. Only those companies guaranteeing the source of the tankage should be used. Additionally, tankage, fish meal, and other animal by-products are not allowed in the diets of certified organic meat animals and some naturally raised systems.

Alfalfa, Clover, Grass, and Succulents

Alfalfa meal is not generally fed as the only source of protein in a pig's diet. Dehydrated meal contains about 20 percent protein, vitamins A and B, and high fiber. Alfalfa meal is not highly palatable to pigs and is usually mixed with other protein sources and grain for a complete feed. Its higher fiber makes it a less desirable supplement for grower and finisher pigs.

Alfalfa pellets can be mixed into a ration for junior and adult pigs as part of the protein supplement. The pellets seem to be more palatable and interesting to the pigs. If alfalfa hay is not available in your area, pellets are an excellent alternative. Additionally, if you are feed-

Overly Fat Pigs

Overly fat pigs are more likely to experience low fertility, farrowing difficulties, lower milk production, more crushed piglets, and increased susceptibility to heat stress, rheumatism, and heart attacks.

Crops grown specifically for hogs can help maintain proper weight. Succulents include mangel beets, sugar beets, turnips, pumpkins, sweet potatoes, and other root vegetables. An area of the garden can be incorporated to plant these vegetables. An added benefit to yourself and the pigs is to let them harvest the root vegetables themselves.

Small or unmarketable potatoes have been a popular feed for swine for hundreds of years. It is believed that cooking the potatoes with a small amount of salt is more nutritious and palatable for the hogs. Potatoes contain little more than carbohydrates, so they will need to be supplemented with protein, vitamins, and minerals. Additionally, green corn or sorghum that has not been frosted or frozen makes excellent succulent feed for pigs.

ing only a couple of pigs, pellet feeds can be very convenient to manage, as they are purchased in 50-pound bags.

Alfalfa and clover hay are usually well liked by pigs. Fed free choice, the pigs utilize the hay for nutrients, fiber, and as a means to occupy their time. These hays are high in protein and can replace pasture in most cases.

Grass hay has little protein but is high in fiber. Older pigs that do not need the additional growth or calories do well on grass hay, but it is not recommended for very small or grower pigs as a primary feed source. Grass hay is also beneficial for helping a pig lose weight. Feeding free choice and limiting the other feed available can help a pig trim down without it feeling too much anxiety.

Kitchen Scraps and Surplus Foods

Pigs appreciate variety in their diet just as much as humans do. Fresh kitchen scraps, vegetable peelings, and surplus vegetables all provide additional nutrients and interest to the pig. Kitchen scraps have been fed to pigs for hundreds of years and can produce some very fine pork. It is the benefit and even the duty of any homesteader to utilize this waste as pig food. Leafy vegetables also add nutrients and variety, but they will need to be accompanied by other more substantial feedstuff. Apples, pears, and other treats can be stored for winter months to add vitamins and variety to the diet. However, a diet high in fruit should be balanced with a protein to avoid diarrhea or weight loss.

Eggs are also a valuable protein source for pigs. Most farms have a surplus at some time during the year. Eggs can be fed, shell and all, mixed in with the ration. Pigs love eggs. In fact, beware of pasturing laying hens with pigs. Once the pigs learn where the treat comes from, they will spend time following your chickens around and gobbling up any eggs they can find to the point that no eggs are left for the farm!

Did You Know?

Pigs like sweets just as humans do. As occasional treats, we have fed day-old bread to our pigs. They definitely showed a preference for cinnamon raisin bread over the onion rolls!

Bakery Waste

Day-old bread and other bakery waste can be fed to hogs with caution. Bakery products are high in carbohydrates and may or may not contain large amounts of fat. Protein content averages about 10 percent. As a carbohydrate, bakery waste can be substituted for corn for up to 50 percent of the ration. Protein, vitamin, and mineral supplementation should be the same as for grain rations. Excess carbohydrates in the diet will create more fat than meat, and prolonged use may produce weakness in the feet and bones. Moistening the bakery products with water, whey, or milk makes them much more appealing to hogs and will help balance the ration.

Fats

Fat is considered a high-energy feed source. Adding fat to the diet of pigs may result in a decrease in feed intake and ultimately fatter carcasses. Fat is only a supplement, not a ration, and cannot be fed as a primary food source. There are times when additional fat may be required, such as when a sow has been depleted from nursing or when an infirm pig has lost a significant amount of weight. Most often, though, a proper diet supplemented with dairy products will be sufficient to restore a pig to prime condition.

Additional Feedstuff

Molasses is typically added to feed as a binder to prevent dustiness and to make unpalatable feeds more sweet tasting and appealing, thus reducing waste. Molasses is also used as a laxative for sows just prior to farrowing. Molasses contains many vital nutrients such as calcium, manganese, copper, iron, and potassium.

Acorns and chestnuts are premium feed for hogs. In fact, in Spain and Italy,

Vitamins and minerals can be found in the soil upon which pigs graze. Pigs will actually eat dirt to gain vital nutrients such as iron and to help balance their nutritional needs.

Assessing Your Pig's Needs

Remember that pigs can be raised in a healthy fashion on a large variety of feeds. You will need to pay attention to your individual pigs to gauge the amount of feed and type of feed that keeps them fit and healthy. Baby pigs will rarely overeat, but adult pigs may continue to stuff themselves if given the opportunity. By feeding two or three times per day, you develop a relationship with your hogs and can better determine their needs. An excessively hungry pig is likely lacking in nutrients and fiber. Try adding alfalfa or clover hay to the feed ration. Feed needs will change with age and sexual activity and during gestation. Refer back to the chart titled Protein Ration and Gain, noting the appropriate level of protein needed for each stage of a pig's life. Carefully evaluating your hogs and adjusting their feed accordingly will keep them in top shape. You can avoid most health problems by balancing the diet and not overfeeding.

respectively, hogs finished on diets of acorns or chestnuts command top dollar, up to $350 for one ham! An orchard or tree stand can be fenced in the fall months, allowing the pigs to clean up anything that may have fallen to the ground.

Pigs can also be fed grass clippings, sunflower seeds, weeds, and leaves as long as they have not been sprayed with pesticides.

Vitamin and Mineral Supplements

Vitamins and minerals are of great importance to the health and development of pigs. If feeding a ground ration, it is best to have a premixed supplement added to your feed. These premixes are formulated based on the age and needs of your pig. If you live in an area that is particularly deficient in some mineral, an additional supplement can be added. Check with your veterinarian or local feed mill for specific deficiencies that might be present in your area.

If your hogs are on pasture, a large portion of their mineral and vitamin needs will be met through grazing. A pastured hog should be offered mineral and vitamin supplements free choice to allow the pig to balance its diet naturally. Forcing pastured pigs to eat minerals and vitamins mixed in to the ration may cause toxicity. Block salt and minerals are not appropriate for hogs, as they will not spend the time licking them as other livestock breeds may do, and some binders used to hold the blocks together can be toxic to pigs.

CHAPTER FIVE

Safe Handling, Routine Care, and Health Issues

Pigs are easy to live with if you spend some time learning their basic needs and behaviors. Routine health care is minimal when a pig herd is well fed, well housed, and well appreciated. Most health issues can be avoided or successfully treated immediately, preserving your herd and its longevity.

Handling

An understanding of pig psychology will save you time and energy when handling pigs. Pigs can be extremely stubborn and will typically do the opposite of what you want them to do. By routinely handling your pigs, they will become accustomed to your care and movements and will be more likely to cooperate when you want to move them about.

Catching Pigs

The shape of a pig's head does not lend itself well to being held. Different methods and equipment are used to hold on to pigs. A hurdle, or pig board, is a lightweight board with handles cut out along the top. Typically about 30 inches high and 2 to 3 feet in length, a pig board can be used to block a pig or direct it to wherever you want the pig to go.

You can also drive pigs into crates for handling. Premade crates are available, but a simple one made of stout boards will serve just as well. Stanchions can be used on pigs, although the shape of the neck may make this difficult.

You can catch small pigs by grabbing a back leg, swinging the pig clear of the ground, and then maneuvering it into a manageable position.

Advice from the Farm

Handling Your Pigs

Our experienced swineherds offer helpful suggestions for handling your pigs.

Know Your Pigs

"Get to know your pigs, and let them get to know you. Talk to your pigs and announce your presence when you are approaching their pens. They will learn to recognize your voice and will feel safe. Don't intentionally make loud noises or upset them, as they will shy away from you the next time you come around. When you are done handling them, give them a bit of food as a reward."

—Bret Kortie, Maveric Heritage Ranch Co.

Move 'Em Out

"Pigs always move better if it is their idea. It takes some practice to be a step ahead and know what the pig is going to do. Be calm and look at it from the pig's point of view. He is two feet off the ground, you are five feet off the ground. Look at his level and see that he has a clear path to go. Move slowly and try to get it the first time."

—Josh Wendland, Wendland Farms

A properly snared pig can be held securely for routine veterinary care, tusk trimming, or blood draws. Hold the snare upward, as seen, and the pig will back away, creating tension on the snare and holding it securely.

Holding Pigs

Holding pigs is an ear-piercing experience for the handler. A set of earplugs will be of great value when a pig must be held or restrained. Pigs should never be suspended by the nose or tail. Small pigs are best held with one hand under the belly and chest and the other pressing on the back. Pigs do not like to be flipped on to their backs or picked up. Avoid carrying small piglets, unless you are training them to be handled in this fashion.

You can catch medium-size pigs by a back leg, then place the pig into the sitting position between your legs while holding on to the front feet. Squeeze the

Hold piglets by placing one hand on the back and one under the stomach as seen here. This allows safe handling of the piglet and makes it feel secure.

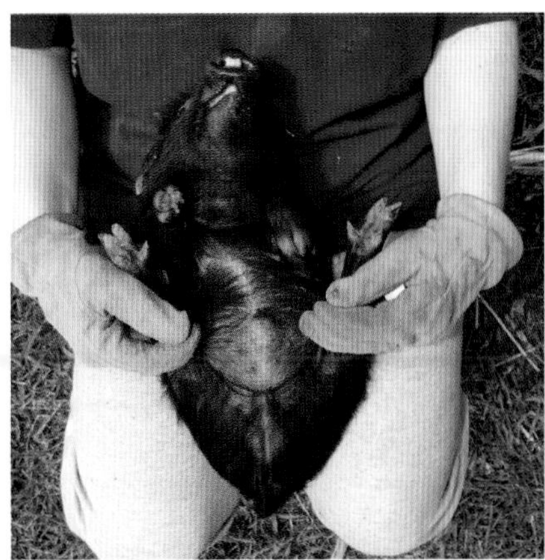

When using two people to castrate small pigs, the holder can lay the pig on its back and pull the back legs forward. This allows easy access for castrating and a secure hold on the piglet.

front feet in toward the head to maintain control of the pig and avoid getting bitten.

Large pigs will need to be snared. A snare is a device that, when placed over the pig's snout or top jaw, can be tightened or manipulated to hold the pig fast. A rigid snare is a metal rod with handles at both ends. Put the handle portion into the pig's mouth, then lift the rod into the air, wedging the snout in the handle. The disadvantage of a rigid snare is that you may take the flesh off the pig's nose. An overly aggressive pig can usually be restrained with a rigid snare.

A cable snare is essentially a noose through a rod that is tightened around the snout. Cable snares have less potential to cause injury to the pig and can be released quickly in an emergency. If a snared pig breaks free from your grip, step back, away from the pig. It will usually flail its head about in an effort to release the snare. A flailing snare can easily break a bone on a human or cause serious injury. After a few tosses of the head, the snare will fall off the pig's snout. You can then pick it up and try snaring again.

Casting or Throwing Pigs

Casting, or throwing, pigs is a method used to lay a pig down and restrain it for treatment, such as administering vaccinations, detusking, or checking an area that you could not reach otherwise. You can easily cast small pigs by picking them up and laying them over on their backs. Cast a larger pig by reaching

Did You Know?

Pigs can scream LOUDLY. The scream of a jet engine taking off measures 113 decibels. The scream of a frightened pig can measure 115 decibels.

Handling Tips for Pig Owners

Here are some helpful tips to use when handling and moving pigs:

- Don't try to make pigs do anything. Try to entice them or outsmart them, as it is unlikely you will be able to outmaneuver them.
- Don't frighten pigs by making loud noises, moving suddenly, or shouting. Move slowly and talk to them, or offer a scratch or a pet.
- Driving and pulling pigs is difficult. It will be much easier on you if you set up panels or gates and crowd the pigs along, giving them only one way to go. You can also coax them with feed.
- Don't try to force pigs into a pen or chute that is dark on the opposite end. Place a light at the end of a tunnel, and they will be more likely to walk in.
- Use a pig slapper, basically a large rattle, to coax pigs along instead of hitting them.

Calmly walking behind a pig will encourage him to move along. Yelling, hitting, and pushing are not necessary and will often result in the pig going away from your intended place.

- Provide yourself with an out if necessary, either over a panel or by working outside the alley or tunnel. Do not allow yourself to be cornered by a pig. A frightened pig may become vicious.

under the pig, grabbing the two opposite legs, and pulling the legs toward you. The pig should fall away from you. Hobbles can be used to secure the hog while you are working on it.

Routine Care

Hoof trimming, detusking, and ringing noses are all practices that the small-scale hog producer should be familiar with. Many herders never detusk boars or trim hooves because of their personal management practices or preferences. If these are practices you wish to maintain in your herd, you can expect to detusk at least once per year and trim hooves up to six times per year, depending on the surface in which your animals spend most of their time. Have a qualified veterinarian perform the procedures the first time and explain them to you, step by step. Although most hobby farmers may never need to perform these tasks, all should know how in case circumstances require.

Trimming Hooves

Some people trim hooves on show pigs to create a well kept and even appearance.

Otherwise, it is usually not necessary to trim hooves. Most pigs raised outdoors get plenty of exercise and continually wear down their hooves. However, pigs raised in very soft bedding or in a confined situation will need their feet trimmed periodically. If the hooves of your pig grow too long or split, you will need to trim the feet to avoid abscesses, infections, and rot.

Splits typically happen on one toe, usually when the ground has become too dry or hard or when a lot of stomping is going on, such as when flies are bothersome. The actual wall of the hoof will develop fissures or cracks that can lead to more serious problems. Split hooves may require the assistance of a veterinarian to properly evaluate the extent of the damage and the possible remedies.

Hire a veterinarian or experienced pig person to teach you how to trim hooves if you have never trimmed before. Serious injury can result from improperly trimmed hooves, so you want to be certain that you understand where and how much to take off the hooves during trimming. Smaller pigs can be held on their backs for hoof trimming, but mature animals will likely need to be cast in a manner mentioned

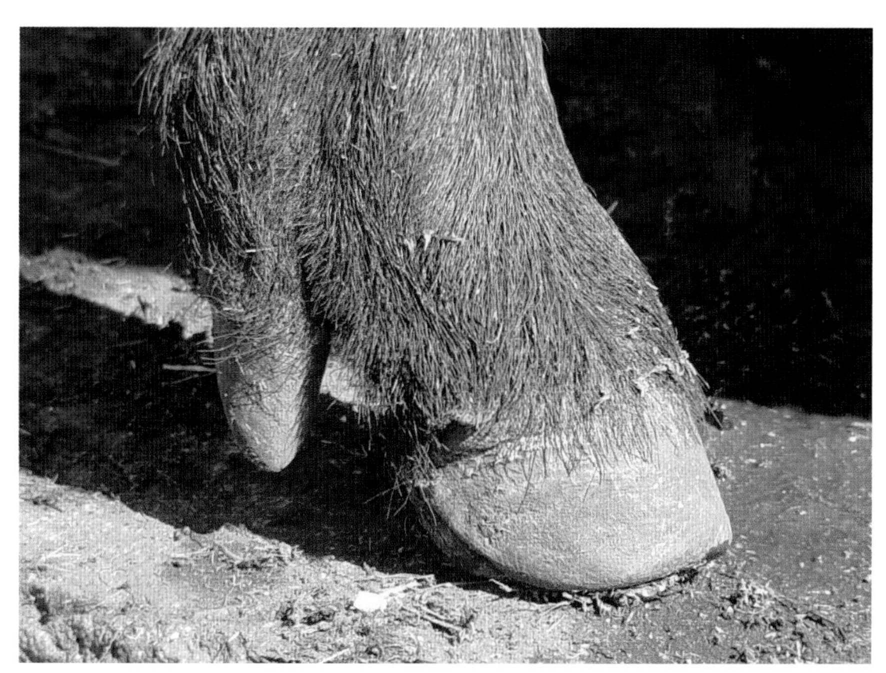

Pigs that are allowed to roam a large area or have access to a small amount of rocky soil or concrete generally wear their hooves down naturally. Mulefoots have a single hoof that wears similarly to cloven-hooved pigs.

earlier. Use a regular hoof knife or pocket knife for the initial trimming. Trim each side by cutting from the back of the hoof toward the tip, so the inner side of the toe is slightly shorter than the outside. Trim with a light hand to avoid cutting into the quick, or soft tissue, which will cause lameness and bleeding. If you accidentally cut into the quick, use blood-stop (styptic) powder, cornstarch, or even a wad of cobwebs to help clot the blood and stop the bleeding. A splash of a copper solution will prevent any bacteria from entering the cut and will firm up the remaining hoof wall. After removing excess hoof with the knife, you can use a common rasp to even up the hooves. When the hooves are trimmed properly, the animal should stand straight up on its toes.

Cutting Tusks

The tusks of boars sometimes grow to such a length as to be dangerous to the handler or to other livestock. Tusks can grow as much as three inches per year, and they should be trimmed annually. Boars constantly sharpen

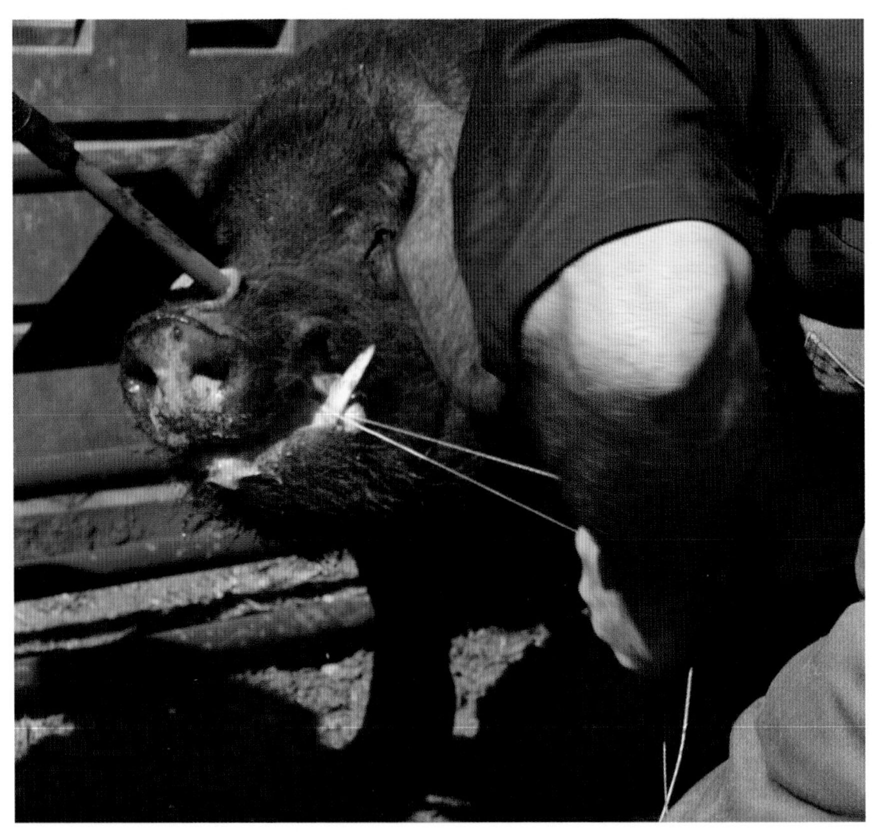

Detusking a boar with a wire saw is a fast method, although it takes a bit of muscle on the part of the handler. Tusks should be trimmed even with the surrounding teeth, taking care not to fracture the remaining portion of the tusk.

their tusks by scraping the top tusk against the bottom tusk, resulting in a sharp, fine point on the lower tusk. Boars that will be used for breeding should be detusked to avoid injury to the sows during the premating introductions and in case the sow rejects the boar. Boars used for show should also have their tusks trimmed.

An experienced swine person or veterinarian should teach you how to perform this task. Detusking a boar is a two-person job. A boar in a snare can still flail his head around and injure a handler, so a strong person is needed to hold the pig.

Use a molar cutter (which was originally designed to trim a horse's teeth), a wire saw, or very sharp bolt cutters for the task. Snub (tie) the boar to a post or snare the boar. Slip the molar cutter over the pig's tusk, and cut off the tusk so that it is even with the surrounding teeth. You can use a wire saw in the same manner to remove the tusk, working the wire back and forth until the tusk comes off.

If you use a bolt cutter, be careful to cut the tusk even with or slightly above the height of the surrounding teeth. Make sure that the bolt cutter is sharp. A dull bolt cutter may actually crush the tusk instead of cutting it and may fracture the tusk down into the gum line. This could cause abscesses, infection, or other serious problems for the boar. If you suspect that you fractured the tusk, an antibiotic may be in order to protect against infection.

Ringing Noses

Pigs use their snouts for smelling, rooting, and digging as well as to maneuver or nose their potential mates. Sometimes it is necessary to ring the nose of a pig to prevent it from excessively rooting good pasture or to prevent it from digging under fences.

Never ring a pig that is less than three months old, as this will retard its growth. A newly ringed pig will need some time to adjust to the pressure the ring puts on its nose and may stop eating for a few days while adjusting to its presence.

Ringing a boar that will be used for breeding purposes is strongly discouraged. A boar uses his nose to stimulate the sow or gilt and to position her for mounting. A ring may cause the boar pain, thereby rendering him unwilling to breed.

There are several sizes and styles of rings, each requiring its own special pincers to set the ring. Choose a ring appropriate to the size of the pig; an overly large ring will quickly be torn out. Insert the rings either in the rim of the snout or through the partition (septum) between the nostrils. One ring is usually enough when set through the septum, although up to three rings may be needed for the rim of the snout. Place one ring in the center of the rim, and place the other two a short distance to either side.

You can hold small pigs in a sitting position for ringing, but larger hogs need to be restrained by a snare or snub. Set the ring quickly and release the pincers right away to avoid tearing the ring out if the pig jerks its head away.

Health Issues

Through proper management and feeding, most illnesses can be avoided. If your pig herd is subject to repeated outbreaks of a particular illness, a thorough consultation with your veterinarian is in order. A vaccination program may be your only defense against repeated illnesses.

Prevention is still the best medicine with pigs. For example, allowing pens and pastures to lie idle for months or, ideally, a year gives some infectious pathogens time to die off before pigs are reintroduced to the area. If it is practical under your management schedule, stagger farrowing, weaning, and other movements of pigs to allow each group to become well established. This will also make it easier for you to identify and isolate problems and provide the needed care and treatment.

Sudden Illnesses or Emergencies

Expertise grows with experience; until you have experience in raising pigs, contact your veterinarian for advice and treatment, especially for sudden illnesses or injuries.

Be prepared to describe all the symptoms your pig is displaying. For example, if your pig has diarrhea, is it watery or pasty? What is its color? Does it have a particularly foul smell? If the pig is displaying any unusual

Ringing noses prevents rooting and some chewing. A ring that has been accidentally torn out of a nose leaves a permanent scar, as seen here.

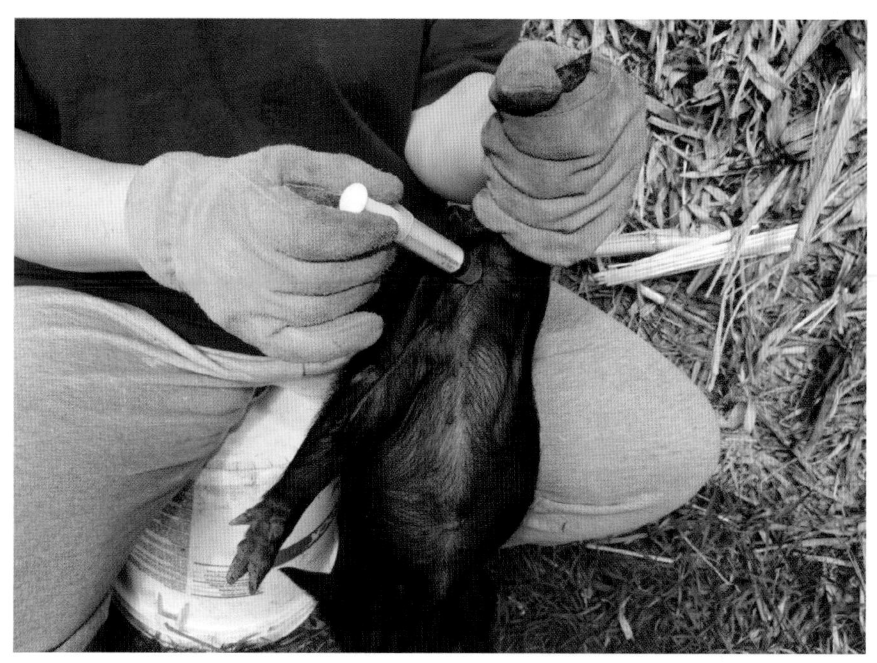

Vaccinating small pigs can be done by one person. Push the skin aside before inserting the needle. Be sure to vaccinate in the fleshy part of the inside thigh. Let the skin slide back after removing the needle to prevent bleeding.

behavior, make note of that also. For example, is the pig paddling, spinning, or backing up? Describe any discharge around the eyes, ears, nose, or mouth. Has the pig recently been weaned, or is it an older pig?

If you suspect that your pig has eaten something poisonous, be certain to tell your veterinarian right away. Immediate action may save your pig.

Vaccinations

In some areas and under certain management practices, vaccines are necessary. Speak with your veterinarian on this point, as he or she may be able to offer advice on diseases or nutritional deficiencies that are common to your area. If you purchased disease-free stock and make no additions, routine vaccinations may not be necessary. If your herd will be taking in new stock or you will be rotating feeder pigs through your farm, then a vaccination program may be in order. Vaccines are available for many pig illnesses, but any illness should be properly diagnosed by your veterinarian before a vaccine is administered. Certain regions and breeds of pigs may be more susceptible to some diseases, while other areas rarely see an outbreak. Avoid vaccinating for illnesses that you are unlikely to encounter. Again, your veterinarian can inform you about the disease threats in your area or make a conclusion based on blood tests from your pigs.

First Aid Kit

All pig owners should have a basic first aid kit for life's little emergencies. Stocking and periodically updating a portable tackle or tool box is a good way to organize first aid supplies and ensure that you have what you need at hand and on short notice. The following supplies are part of a basic kit:

- A marking crayon or paint
- A razor blade or scalpel
- A rectal thermometer used for babies
- A snare
- An antibiotic powder that can be added to the water supply
- Antibiotic liquid to be given orally
- Assortment of hypodermic needles and syringes
- Betadine solution, hydrogen peroxide, and triple antibiotic cream
- Coppertox (for treating foot injuries)
- Electrolytes or instant sport drink mix
- Eye dropper or syringe for orally dosing piglets
- Fly spray
- Obstetric lubricant
- Pepto-Bismol
- The phone number for your veterinarian
- Prescription medications as recommended by your veterinarian
- Scours (diarrhea) remedy that can be added to the water supply
- Spray bottle with Listerine
- Veterinary manual for pigs
- Wound spray

Ask your vet to teach you how to give shots yourself, as this will save you much in costs and time. Being able to vaccinate or treat your own pig will make you a better manager of your herd. You will no longer need to wait for the vet to arrive, and you will be able to vaccinate as time permits.

Worming and Internal Parasites

Preventing and treating parasites can be a challenge at first, but with proper management and preventive worming practices, the bulk of parasites can be eliminated from your herd of pigs. You will never have a pig herd that is completely free of parasites. Reinfestation is common and easy, as pigs constantly have their noses on the ground. The objective is to keep the parasites at a minimal level.

Prevention

Your chosen management system is the determining factor in the parasite load carried by your pigs. It is a good practice to isolate and worm all incoming pigs to your herd. Feces from quarantined animals must be removed and disposed of in an area that your pigs do not have access to.

Internal parasites are not a common problem in weanling, growing, and adult pigs unless the pigs are housed in continuously populated pens or stalls. Contaminated bedding harbors parasites, as can concrete floors. Proper disinfection of pens and stalls between groups of pigs will eliminate some of these problems.

Pigs raised on dirt may also be more susceptible to parasites if the dirt is allowed to be continuously covered in feces and urine, with no rest between groups of pigs. Allowing a particular pen or pasture area to rest for several months between occupants will also eliminate many parasites. This is particularly effective in summer, as UV rays from sunlight and the heating up of the soil will kill many pathogens. Moist or wet areas encourage parasite reproduction, as does an area that is never exposed to sunlight.

If you are raising your pigs on pasture or dirt, worming should be performed at least every six months to ensure that the parasite load is not building up in your soil. When planning to move outdoor pigs to fresh paddocks that have previously not been used for swine, worm the group of animals fourteen days prior to the move. Moving outdoor sites every year or two years will also help break the life cycle of parasites.

Keeping your sows free of parasites will not only keep your sows in top order but will also prevent the spread of parasites to piglets. Most parasites are housed in the feces of the adult pigs, and piglets can easily become infected while nosing around behind the sows. Piglets should be wormed at weaning to rid them of any parasites that may have been transmitted by the sow. An additional worming performed two weeks later will also prevent any further spread of parasites. Then, when weaning piglets, move them to fresh ground or newly disinfected pens for additional parasite prevention.

Feces should not be allowed to remain in pens for more than three to four days, as this time frame allows parasite eggs to grow and hatch. In an indoor system, an "all-in, all-out" cleaning regime will prevent a toxic buildup of parasites. This does not mean that the pens should not be properly disinfected between groups, however. Some people recommend burning all old bedding to prevent the spread of parasites and disease, and this may be an option that is appealing to you. Composting will not necessarily kill all parasites.

Symptoms and Treatment of Parasites

Internal and external parasites subsist on the swine host. They are found in the digestive tract, kidneys, liver, lungs, blood stream, and ears and on the skin. Symptoms of parasitic infection include coughing, loss of body condition or weight, bald or flaky skin patches, vomiting, anemia, diarrhea, blood in feces, joint pain or stiffness, pneumonia, and even death.

In most cases, you will not be able to see the parasites in your pigs. Unless the pigs are heavily infested and actually shedding the worms in the feces, you will not be able to determine a parasite type by purely visual methods. Parasite fecal exams should be performed by your veterinarian or medical laboratory. By testing periodically, you can be certain that you are not increasing your parasite load and will also be able to determine the proper treatment, if necessary.

Medicines used to treat parasites are called anthelmintics, commonly known as wormers. These can be dispensed in the water or feed, poured on the skin, or given through injections or orally. Pour-on wormers tend to be the least effective, as the skin on a pig is generally tough and does not absorb enough of the product to be effective.

Dosage is usually determined by the weight and age of the animal. If a parasite load is a problem for your herd, a rotating worming program is in order. This program would require you to worm your animals as much as once per month and to rotate the wormer used. For example, there are four common wormer products used in swine: Ivermectin, Doramectin, Fenbendazole, and Levamisole. These worming products cover a broad spectrum of parasite infestations. By rotating the use of these products, you will prevent the parasites from building up a resistance to the products and eliminate a higher percentage of the parasites after each worming.

Once your herd has had a significant reduction in parasite load, and assuming that the pens and housing areas are kept clean of feces, you can reduce your worming schedule to once every six months. Any program, however, should be thoroughly discussed with your veterinarian.

Wash Your Hands

Although most parasites and diseases are particular to each species of mammal, some may be transmitted between animals and humans. These are called zoonotic diseases. Trichinosis, tapeworm, and the fungus ringworm, for example, can be transferred from pigs to people.

To reduce your risk of infection, wash your hands thoroughly after handling pigs. Be careful not to transfer pig feces into your living areas by tracking it in on your shoes or clothing. Biting your nails is also a way to transfer worms, fungus, and bacteria to humans. Children tend to put their hands in their mouths, so care should be taken to thoroughly wash their hands whenever children are exposed to anything that is soiled by pigs.

Common worms found in pigs include ascarids, kidney worms, lung worms, muscle worms, nodular worms, pork bladder worms (human tapeworms), red stomach worms, round worms, stomach hair worms, thick stomach worms, thorny headed worms, thread worms, and whip worms. Additional parasites include lice, mange (mites), coccidia, the protozoan *Toxoplasma* (cause of toxoplasmosis), cryptosporidia, and several bacteria.

For more information regarding swine parasites, see http://www.thepigsite.com. There you will find details, photographs, and recommended treatment methods.

Scours

Several bacteria, viruses, illnesses, and diseases may cause pigs to scour (have diarrhea). Piglets are particularly at risk when scours develops, as they can quickly dehydrate and starve. An inexperienced hog person should contact a veterinarian immediately at the first sign of scours. Determining the cause will be your first step in curing the pig and preventing further episodes.

Adding electrolytes to the water source to bolster the pig's electrolyte balance during increased fluid loss will help stabilize the pig while you determine the cause and begin the proper treatment. Electrolytes come in powdered form in small pouches and can be purchased at most farm stores. In a pinch, a bottle of sports drink will also provide the needed nutrients and replace fluids, and the sweet taste may encourage your pig to drink.

Conclusion

The health of your herd and regular interactions with your pigs are the major contributing factors to the satisfaction you receive while raising them. With these simple, routine health management strategies and proper attention to the changing needs of your pigs, you should realize a long-term relationship with your animals and enjoy a productive level of husbandry.

CHAPTER SIX

Pig Breeding and Farrowing

Breeding pigs can be quite challenging, especially to the first-time pig breeder. Although raising your own piglets is one of the highlights of pig husbandry, it may be best to start with previously weaned piglets and raise them strictly for freezer meat. Alternately, purchasing an already proven sow may be a more manageable method the first time. This will give you some experience in handling the animals, and you will become more familiar with the behavior of pigs.

If you are contemplating keeping a breeding herd, including boars, sows, and growing pigs, keep in mind that a sow eats approximately 2,000 pounds of feed per year and may not breed on schedule or perhaps not breed at all. Then consider the boar, which eats even more and complicates your management system by needing its own pen and other special concessions even when it is not in service.

Then look at the averages in piglet rearing. Producers lose, on average, about 25 percent of all live born piglets, most within the first four days of life. Piglets are susceptible to many diseases and health issues during their first eight weeks of life, further complicating your management practices. Mothering ability varies greatly among sows, and you will not be able to predict whether the sow will lie on piglets, savage the babies, not provide enough milk, or just refuse to take care of the piglets.

Breeding and raising your own pigs may sound like a dismal proposition, and, in fact, many things can go wrong. However, it is a very rewarding part of farming. The skills of swine breeding can be acquired over time, and attention to detail will improve your chances of raising healthy piglets to adulthood.

Newborn piglets under four days old are vulnerable to a variety of problems, including being stepped on, lain on, and being exposed to cold and moisture. Extra care should be taken to see that your piglets are safe and warm and ensure that they make it past this critical time.

If you decide to rear your own piglets, prepare yourself as much as possible by talking with other breeders and reading any information you can find on farrowing. Classes are sometimes offered at colleges, at extension offices, and through artificial insemination companies. Ideally, a good sow will do all the work, and you will have to do little more than look on in admiration.

Did You Know?

The world's largest litter of pigs was thirty-four piglets, born to a sow in Denmark in 1961. Labor took two days. Fortunately for the mothers, most pigs have only eight to twelve babies in a litter.

Selection

Selecting breeding stock requires as much knowledge and expertise as is possible. Learn as much as you can before you go out looking. This includes talking to experienced pig people, reviewing the breed standard for the breed of your choice, and looking at pictures of pigs who have won in some category of their breed standard. You will need to evaluate each potential candidate for conformation to the breed standard and disposition and review any family progeny records available.

Sows and Gilts

Evaluate proven sows based on their age and productivity. Examine their progeny records with an eye for the number of pigs weaned in relation to the number of pigs born. Just because a sow has given birth to twelve piglets doesn't mean she is a good mother. How many of the twelve has she weaned, and are all the weanlings healthy and of good weight for their age? A sow that births twelve and raises two is not nearly as valuable as one that births six and raises six.

Gilts are young female pigs that have not had a litter, so they cannot be evaluated on progeny records. Gilts should be chosen according to their conformation and size. Overly masculine traits, such as coarse head, large shoulders, large hams, and burly overall appearance, should be avoided. Count the teats to be certain that the gilt has at least ten, well-shaped, equally spaced teats for nursing. Avoid those with deformities such as inverted teats or needle teats and teats that are randomly spaced.

A red wattle sow enjoys some fresh air and sunshine with her recently born litter. Mama and piglets will return to the hut when the temperature becomes cooler.

Sometimes piglets and sows need a little help getting started. Placing the piglets on the teats will ensure that they get colostrum, which is vital for newborns as it provides energy, fat, and most importantly, antibodies to disease.

If you intend to breed her right away, the gilt should be of sufficient age to breed, usually about eight months. If available, check the gilt's mother's records to find out the litter size she came from and what type of mother she had. Sows and gilts should have a refined, feminine appearance.

Boars

Boars determine the success or failure of your breeding program more than any other factor. Usually, a boar is used on all the breeding females in a small herd, thereby making all the offspring susceptible to any problems he may have. Evaluate a boar by looking at his reproductive organs. The testes should be large and well formed, and the penile sheath

A well-proportioned boar with good body conformation should be your first choice in boar selection. Choose a boar that exhibits the traits you want to see in your herd such as large hams, strong bone, solid feet, and good disposition.

should be close to the body. A boar should also have at least ten well-formed teats, as this trait will pass to his daughters. Boars should have a masculine appearance and be athletic and active. Legs must be thick and stout to be able to carry the boar and support his weight during breeding. Large hams will help him support himself while mounting females.

The Breeding

Get your breeding area set up ahead of time. Have an easy method of moving the sow to the boar, such as an alleyway directly from one to the other or a pig carrier to move the sow. Allow yourself ample room to be out of the way during mating. Be prepared to remove the sow quickly if any serious fights occur or in the event of excessively aggressive behavior on the part of either pig.

Introductions

Introducing a young boar to a breeding group can be dangerous for the boar. Do

Artificial Insemination

Artificial insemination (AI) of pigs has become increasingly popular in recent years. Deep-freezing and storing semen are still imperfect sciences, so fresh semen is typically used. Semen can be shipped from many private farms or sperm companies, but obtaining it from a local source may be the most cost effective. AI may be very useful to farmers with too few sows to justify maintaining a boar, and closed herds can use AI to bring in new genetics without the risk of introducing disease.

Artificial insemination will not be as successful as natural breeding if it is not performed properly. Poor results may be due, in part, to poor timing of ovulation, poor quality of semen, or improper handling or hygiene of the semen or equipment. Costs for straws of semen can be high, especially when semen is from boars of champion bloodlines or with high performance records. If, however, only a few sows are to be bred, AI may actually cost less than the price to maintain a boar year round.

A variety of colors and body shapes can be seen when crossbreeding pigs. Selecting for excellent qualities, no matter what the breed, garners the best offspring.

not put your young boar in a pen of several females of various ages and sizes and expect him to fend for himself. Females may attack the boar or rough him up, causing injury not only to his body but also to his pride. A chastised boar may be hesitant to breed for several weeks or months, setting back your program and adding more costs.

It is best to start a young boar with one female at a time, allowing him to learn the ropes. There is something to be said for pairing an experienced sow with an inexperienced boar, but they should be matched fairly close in size to avoid any problems.

The same applies to gilts being bred for the first time. Size does matter, and mating a very large boar with a small gilt may be disastrous. Small females may lie down and refuse to stand to be mounted. The weight of the boar may cause injury to the female's back. Older, experienced boars don't take the time to find out if the female knows what to do and will mount regardless. Some older boars are rough with their mates, so you should watch carefully to be sure that the boar does not tusk, bite, or otherwise injure the female.

Breeding pairs can be introduced through a strong fence to minimize fighting. If you are not sure whether your female is in heat, introducing her to a male through a fence is a good method to start with. If the female backs up to the fence and allows the boar to nose her, it is likely that she is ready to mate. A female ready to mate also stands rigid, referred to as standing heat. You can

Signs of Heat

A female goes into heat every twenty-one days, typically displaying the following signs:

- Swollen or red vulva
- Sticky vaginal discharge
- Excessive grunting or squealing
- Mounting or being mounted by pen mates
- Nervousness
- Standing heat

apply a fair amount of pressure to her back without her moving a muscle to shake you off. This is an almost sure sign that she is ready to mate.

Hand Versus Pasture Breeding

In hand breeding, the sow is brought to the boar for mating and then moved back to her own pen immediately after the boar is finished mounting. Mating takes place within a few minutes if the sow is ready. If she squeals, runs, and tries to get away from the boar, you may have misjudged her cycle. Put her back in her own pen to avoid injury. Once exposed to the boar, a sow may come in to heat quickly as her hormones kick in, so bring the sow back twelve hours later to try again.

Pasture breeding is by far the easiest method for the handler. If you have an experienced boar and sows, you can safely put them together in groups on the pasture, and they will do what comes naturally. The only drawback to pasture breeding is being able to time your farrowing. If you do not see the actual mating, you may not know that it has taken place.

The Mating

Mature boars should be able to service two sows per day, up to forty per month. Junior boars can service one sow per day, up to twenty-five per month. When pen or group breeding, sows should be limited to six to eight per mature boar and four to six per junior boar.

Allow your boar to remain with the sows through only two heat cycles. If they do not successfully mate in that time, you will need to evaluate your pigs to determine the problem. By allowing the boar to remain through only two cycles, you will also narrow

Boars produce foam at the mouth when they are interested in a female. This is part of the mating ritual.

Advice from the Farm

Choosing the Right Sow

Our experts offer valuable information about selection in sows.

Teat Ridge

"When choosing a sow, look for a good underline. The sow should have 12 to 14 functional teats that are evenly spaced. The teat ridge should be under the middle of each side, and the teats should hang straight down. A sow with teats that are angled out will not be able to nurse well, as the lower side of the teats will wind up underneath the body. The piglets won't have access to all the teats, and the stronger ones will get all the milk."

—Jeremy Peters, Madison, South Dakota

Note: Many heritage hogs have only ten teats. This should not be seen as a defect.

Beware Circling Sows

"Sows must have excellent mothering skills. They need to talk to their piglets and remain lying down when farrowing. Sows that get up and circle around are going to crush piglets. Don't keep a circling sow."

—Al Hoefling, Hoefling Family Farms

Wean Enough Piglets

"In a small herd, a sow should give you two litters per year and wean at least six piglets each time. She needs to be able to raise them to weaning without additional supplement above what the rest of the herd gets. A sow should look around and know where her piglets are before she lies down."

—Josh Wendland, Wendland Farms

Cull Bad Sows

"Culling bad sows is hard to do, but all successful breeders cull heavily for bad traits. Your sows define your success. Don't keep a sow that eats piglets, abandons the nest, or lies on her babies."

—Bret Kortie, Maveric Heritage Ranch Co.

Temperature

The weather has some impact on the breeding of pigs. Boars are less likely to breed in very hot weather. Excessive heat actually reduces sperm count, and the animals are less active during hot weather. Limiting breeding to the cooler parts of the day or evening may increase fertility. High heat can also have adverse effects on sows, as heat can promote the loss of embryos.

By the same token, extremely cold weather can inhibit a pig's fertility. During cold weather, pigs utilize feed to maintain their body temperature. If sows are expected to perform under very cold conditions, their feed must be increased to accommodate the loss of body heat and the extra activity during breeding.

the farrowing time to fewer than thirty days. Remember that a heat cycle happens every twenty-one days and can last from two to five days per cycle. By limiting the mating opportunities to two cycles, you will better be able to predict farrowing dates. Any pigs that do not breed within the two-cycle time frame should be evaluated for health issues and culling.

If you are maintaining more than one boar and you are breeding piglets strictly for meat production or as terminal replacement animals, you may benefit from placing one boar with the sows in the morning, and another boar in the evening. This may increase their fertility and litter size. This practice cannot be utilized for purebred herds, however, as all litters are recorded with the actual parents of the piglets. If mating with two boars, it would be impossible to tell which sire was responsible without a DNA test.

Gestation

Pig pregnancy lasts for approximately 3 months, 3 weeks, and 3 days, an average of 115 days. Up until about 80 days, your sow will show no visible signs of pregnancy. Portable ultrasound can be used to determine whether the pig is pregnant.

Feeding the Gestating Sow

During the first two-thirds of her pregnancy, a sow will do well on a maintenance ration with a protein content of about 14 percent. Gilts and very thin sows need to be fed additional feed or a feed with higher fat content for them to continue to gain weight and adequately provide for the piglets they are carrying.

The fetal piglets make their surge in growth during the last third of gesta-

Sows in the late stages of pregnancy start to develop a teat ridge and a slight protrusion in the abdomen just behind the ribs. Most sows do not show signs of pregnancy until the last few weeks of gestation.

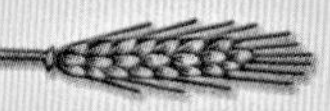

114-Day Swine Gestation Table

The following chart can be used to determine the likely farrowing date of your sow. For example, say the sow was bred on January 10. Look directly below January 10 on the chart, and you will see a farrowing date of April 4. This is an approximate date. You should be prepared for farrowing at least five days prior and up to two weeks after the projected due date.

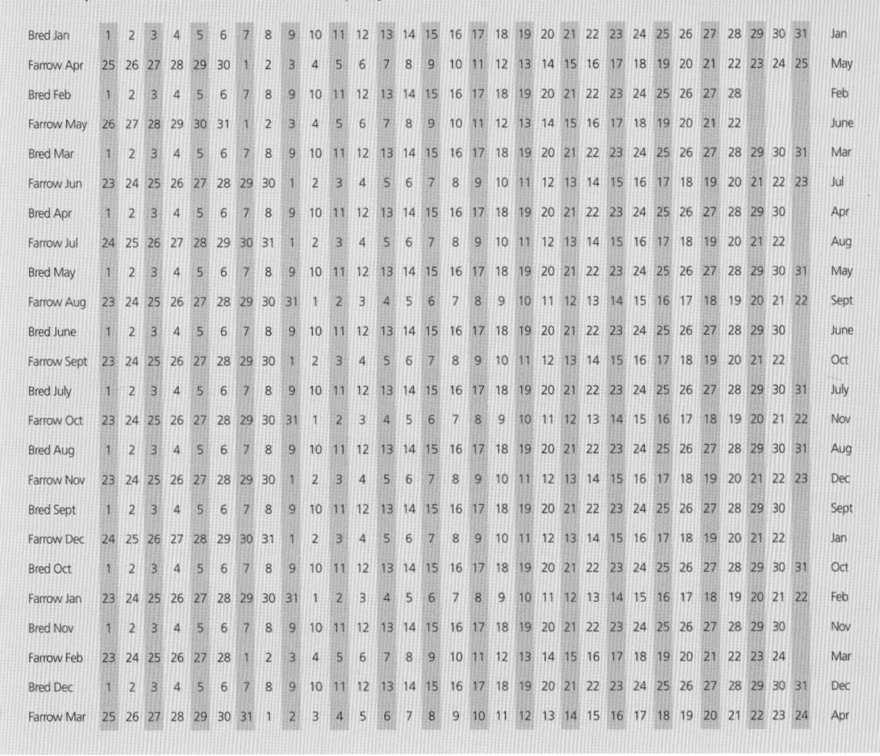

Bred Jan	1	2	3	4	5	6	7	8	9	10	11	12	13	14	15	16	17	18	19	20	21	22	23	24	25	26	27	28	29	30	31	Jan
Farrow Apr	25	26	27	28	29	30	1	2	3	4	5	6	7	8	9	10	11	12	13	14	15	16	17	18	19	20	21	22	23	24	25	May
Bred Feb	1	2	3	4	5	6	7	8	9	10	11	12	13	14	15	16	17	18	19	20	21	22	23	24	25	26	27	28				Feb
Farrow May	26	27	28	29	30	31	1	2	3	4	5	6	7	8	9	10	11	12	13	14	15	16	17	18	19	20	21	22				June
Bred Mar	1	2	3	4	5	6	7	8	9	10	11	12	13	14	15	16	17	18	19	20	21	22	23	24	25	26	27	28	29	30	31	Mar
Farrow Jun	23	24	25	26	27	28	29	30	1	2	3	4	5	6	7	8	9	10	11	12	13	14	15	16	17	18	19	20	21	22	23	Jul
Bred Apr	1	2	3	4	5	6	7	8	9	10	11	12	13	14	15	16	17	18	19	20	21	22	23	24	25	26	27	28	29	30		Apr
Farrow Jul	24	25	26	27	28	29	30	31	1	2	3	4	5	6	7	8	9	10	11	12	13	14	15	16	17	18	19	20	21	22		Aug
Bred May	1	2	3	4	5	6	7	8	9	10	11	12	13	14	15	16	17	18	19	20	21	22	23	24	25	26	27	28	29	30	31	May
Farrow Aug	23	24	25	26	27	28	29	30	31	1	2	3	4	5	6	7	8	9	10	11	12	13	14	15	16	17	18	19	20	21	22	Sept
Bred June	1	2	3	4	5	6	7	8	9	10	11	12	13	14	15	16	17	18	19	20	21	22	23	24	25	26	27	28	29	30		June
Farrow Sept	23	24	25	26	27	28	29	30	1	2	3	4	5	6	7	8	9	10	11	12	13	14	15	16	17	18	19	20	21	22		Oct
Bred July	1	2	3	4	5	6	7	8	9	10	11	12	13	14	15	16	17	18	19	20	21	22	23	24	25	26	27	28	29	30	31	July
Farrow Oct	23	24	25	26	27	28	29	30	31	1	2	3	4	5	6	7	8	9	10	11	12	13	14	15	16	17	18	19	20	21	22	Nov
Bred Aug	1	2	3	4	5	6	7	8	9	10	11	12	13	14	15	16	17	18	19	20	21	22	23	24	25	26	27	28	29	30	31	Aug
Farrow Nov	23	24	25	26	27	28	29	30	1	2	3	4	5	6	7	8	9	10	11	12	13	14	15	16	17	18	19	20	21	22	23	Dec
Bred Sept	1	2	3	4	5	6	7	8	9	10	11	12	13	14	15	16	17	18	19	20	21	22	23	24	25	26	27	28	29	30		Sept
Farrow Dec	24	25	26	27	28	29	30	31	1	2	3	4	5	6	7	8	9	10	11	12	13	14	15	16	17	18	19	20	21	22		Jan
Bred Oct	1	2	3	4	5	6	7	8	9	10	11	12	13	14	15	16	17	18	19	20	21	22	23	24	25	26	27	28	29	30	31	Oct
Farrow Jan	23	24	25	26	27	28	29	30	31	1	2	3	4	5	6	7	8	9	10	11	12	13	14	15	16	17	18	19	20	21	22	Feb
Bred Nov	1	2	3	4	5	6	7	8	9	10	11	12	13	14	15	16	17	18	19	20	21	22	23	24	25	26	27	28	29	30		Nov
Farrow Feb	23	24	25	26	27	28	1	2	3	4	5	6	7	8	9	10	11	12	13	14	15	16	17	18	19	20	21	22	23	24		Mar
Bred Dec	1	2	3	4	5	6	7	8	9	10	11	12	13	14	15	16	17	18	19	20	21	22	23	24	25	26	27	28	29	30	31	Dec
Farrow Mar	25	26	27	28	29	30	31	1	2	3	4	5	6	7	8	9	10	11	12	13	14	15	16	17	18	19	20	21	22	23	24	Apr

tion (roughly, the last five weeks). These extra demands on the sow will continue throughout lactation. Increasing the feed by 25 to 50 percent is needed to maintain the sow's weight and provide adequate nutrients for the growing piglets.

PREFARROWING

About four weeks prior to farrowing, your sows and gilts should be treated for any internal or external parasites. Review your vaccination program to see whether any animals need to have booster shots.

Get all your farrowing pens in order, but don't add bedding yet. Add bedding just prior to moving the pig in to prevent dust accumulation and contamination from other animals, such as barn cats. Make sure your farrowing pens are cleaned, disinfected, and well lit. Have your feeders and water dishes ready to move in.

Farrowing

Several signs can foretell the coming of the piglets (farrowing), but not all sows display these signs. During the last three weeks of gestation, sows typically start to show a round protrusion just behind the rib cage. A teat ridge will develop, indicating that the udder is filling up with milk. The teats will start to stiffen and protrude. You may notice a reddening of the teats or that they feel warm to the touch.

Readying the Farrowing Area

Just before moving the sow into the farrowing area, place the proper bedding in the pens, and set the water and feed pans. Bedding should not be more than a few inches thick: piglets will burrow into extra thick bedding, making it easy for the sow to lie on or trample the piglets. Bedding should be of a material that is mildly coarse with small particle sizes. Wood chips are ideal for the first few days. When piglets are becom-

Groups of sows can farrow in the same area, provided that each is given her own hut or stall in which to have her piglets. Cross mothering can take place under this farrowing system, providing a better chance for all piglets to be nursed properly.

ing mobile (about three to four days) and have learned to get out of the way of the lying down sow, straw may be added to allow for extra bedding, drainage of wastes, and some material to burrow in.

Move your sows and gilts into the farrowing quarters about one week to four days prior to their due date. This allows them to settle in to their new house and to get accustomed to the feeding and schedule of the new facility. Moving a sow too close to her farrowing time may increase a loss in piglets. The sow may be anxious and distraught over her new surroundings or afraid when separated from her usual pen mates. If you are farrowing several sows that are normally housed together, try to place them near each other in the farrowing facility. This will also reduce their anxiety. An anxious or nervous sow will spend more time getting up and down during the farrowing, and will be more likely to crush her piglets or step on them.

Allow the sow to move the bedding and pile it as she sees fit; this behavior is a manifestation of a natural nesting instinct. Sows usually build nests two days to a few hours before farrowing, another indication that they are getting ready.

Prefarrowing Ration

Approximately one week prior to farrowing, the sow's ration should be adjusted. The farrowing ration is typically a 14 percent protein ration with an added laxative such as molasses. The laxative softens the stool without dehydrating the sow and reduces milk production for the first three days. A reduced milk production prevents the piglets from scouring by overeating, and it prevents udder edema and milk caking on the udder.

The Sow's Behavior

The sow will become restless just prior to farrowing. She may get up and down several times, trying to position herself for the birth. Try not to disturb her while she is settling in or during the actual birth process. By getting up and down, the sow may mash her piglets. Once lain on, piglets rarely survive.

A sow in labor appears to be in a meditative state. She will roll up on her side and extend all four legs outward. During contractions, she will pull the back legs forward and against the stomach.

Once farrowing starts, the piglets should come along at about twenty-minute intervals. Some people believe that the smallest pigs are born first and last. Piglets can be born head or feet first, and either way is normal. Piglets are typically born with the umbilicus still attached to their bellies. The umbilicus is highly elastic and stretches to great lengths. There are also several weak spots along the cord that allow it to break away from the piglet's body, freeing it from the placenta. Unless the cord is excessively long and interferes with the piglet's movement, it does not need to be removed. An extra long

Advice from the Farm

Farrowing

A bit of advice on sow behavior and farrowing preparation from our experts.

Nesting

"A good sow will start to build her nest 48 to 24 hours before farrowing. She will build a nest that is a mound, not a hole. A mound will help prevent mashing the piglets. A hole will have sloped sides, and the piglets will have no choice but to roll under the sow. Good mothers talk to their piglets, and they do not get up during farrowing or shortly after. Poor mothers will get up and down, and circle constantly, stepping on and crushing piglets."

—Al Hoefling, Hoefling Family Farms

Breed at the Right Age

"Gilts should be at least 7 months before getting bred. They don't grow much after they've farrowed, so breeding a young gilt will result in a small sow. Their capacity will be impaired by their size, and they'll have small pigs. The best litters come from a 3–4 year old sow; that's when they hit their prime. When gilts have their first litters, they don't milk all that well, so you should be more forgiving of a gilt."

—Josh Wendland, Wendland Farms

Leave Mom Alone

"Heritage breeds tend to have excellent instinctual habits. Give them a clean, healthy environment and they will do the rest. The less you interfere, the better the piglets will be."

—Bret Kortie, Maveric Heritage Ranch Co.

umbilicus should be removed by grasping it at about one inch from the body and at one inch farther down, then pulling the two sections apart. Some herdsman cut the cord to a uniform length and dip the umbilicus in a weak iodine solution. This is not as critical to the health of the piglet as it may be in other livestock. If you are able to handle the piglets as they are being born, clipping and dipping is a good precautionary measure against infection. A clean, dry environment is the best way to avoid umbilicus contamination.

After all the piglets are born, the sow will pass the afterbirth, or placenta. A sow generally eats the afterbirth if it is not removed. Removal is recommended. Some believe that eating the placenta will encourage cannibalism in sows, and most commercial facilities remove the afterbirth to prevent the spread of disease and for hygiene benefits to the piglets.

Aiding the Birth

Sows generally farrow without trouble. If the sow has labored for more than thirty to forty minutes without producing a piglet, an intervention is likely needed. If you have never assisted in a birth before, you should contact your veterinarian immediately. Your sow may have a large piglet stuck in the birth canal or a stillbirth, or she may have just exhausted herself beyond the point of continuing.

If you are assisting the sow, clean your hand and arm thoroughly and lubricate it well with an obstetric lube from your first aid kit. A mild antibacterial soap will suffice in a pinch. Clump your fingers together in a bird beak shape, and slowly enter the vulva. Gently push forward until an obstruction is felt. The birth canal is very small, without much room for maneuvering. If you feel an obstruction, try to identify which end of the piglet you have.

If presenting head first, gently grasp the head and pull toward you at a downward angle. If presenting back feet first, grasp both back feet together and pull toward you. Do this very slowly; do not yank the piglet out.

Once you have removed the obstruction, the birth process should resume, and the placenta should be passed after the last piglet. A retained placenta can cause a severe infection. A shot of oxytocin, which causes uterine contractions, should help to shed the placenta if the sow is not passing it naturally. Check with your veterinarian for direction in proper procedure if you have not had experience with this drug. You may need to apply pressure to the placenta in conjunction with the sow's contractions to remove it. Do not yank the placenta out, as that may tear or injure the uterus.

If your sow had serious birthing problems, you should talk to your veterinarian regarding the causes and remedies. For example, an overly fat sow will have more problems birthing, and you may have to initiate weight management practices to avoid repeating the problem. If, however, the sow had problems of a genetic or dispositional nature, you should consider culling her from your herd. A very large percentage of sows can and will farrow on their own with no trouble. These are the ones to keep.

Postfarrowing Watch List

Once your sow has farrowed, monitor not only the piglets but also the sow's demeanor and performance. A lack of appetite, constipation, low milk production (hypogalactia), or other sicknesses are indications that your sow may not perform her mothering duties well, and you will lose piglets. If the sow seems constipated in spite of the use of

Antibiotics After Birth

If you had to assist the sow during delivery, you may have introduced any number of bacteria to the womb. An injection of antibiotics such as LA-200 or penicillin is in order.

a farrowing ration, give her an injectable laxative or add Epsom salts to her ration at the rate recommended on the package—typically administered by the weight of the patient. Constipation may cause straining, lack of appetite, and discomfort, which in turn will cause a sow to stop nursing. Loss of appetite can severely inhibit milk production, so encourage the sow to eat every time feed is offered. Adjusting the ration to a more palatable mix may help. If your sow has a favorite food such as greens or bananas, add a small amount to the ration to encourage eating.

Failure to produce enough milk to feed her litter is a serious problem in a sow. If the temperatures are sufficiently high, a sow may be too uncomfortable or dehydrated to produce milk. Reduce the temperature in the farrowing area through a cooling pool of water, a sprinkler, or a fan.

The Newborn Piglets

Piglets are born wet, with a thin membrane covering their bodies. This membrane is designed to come off the piglet easily, and in most cases it will do so on its own. Some breeders like to dry the piglet. Although drying is not necessary, it will not harm the little one. A warm environment or a heat lamp will dry the piglet in short order. If the piglet has an excessive amount of the membrane on its face, it is best to clear it away to prevent suffocation.

Piglets will find their way to the teats within a few minutes. All piglets should nurse within the first hour of birth in order to gain strength and get off to a good start.

Watch your newborn piglets carefully to see that all get a belly full of the first milk, or colostrum. Colostrum is vitally important to the health of the piglet, as it is high in fat (energy) and antibodies that provide natural immunity from disease. A piglet will establish itself on a particular teat, and will continue to go back to that teat for the duration of the nursing process. Because of this habit, it is important to keep a watchful eye on the sow's teats to be certain that some do not dry up or develop mastitis. Once established, the piglet will continue to suck the same teat even though it no longer produces milk. If this happens, the piglet will lose weight and eventually starve to death. If a piglet begins to lose weight, it is best to move it to another pen for special feeding or to graft the piglet onto another sow for nursing.

Conclusion

By starting with the best breeding stock available, you will increase your chances of producing large, healthy litters of piglets. Healthy sows have healthy piglets if they are provided with the proper heat, nutrients, and a clean environment.

CHAPTER SEVEN

Care of Piglets

Although most sows do a fine job of raising piglets to weaning age, your husbandry methods affect many aspects of the piglets' lives. What you do to maintain the comfort level of the sow—adequate heating or cooling of the facility, the cleanliness of the facility, and the proper feed supply—plays an important role in the health and well-being of the piglets.

Nutrients for Sows and Piglets

Sows need additional vitamins and minerals during gestation and nursing. By properly feeding the sow, you are properly feeding your piglets. The best diet for a sow does not provide all the nutrients your piglets will need, however. Milk does not contain enough iron to create red blood cells in the piglet and must be supplemented through shots or access to iron-rich dirt. Many farmers who raise their pigs on dirt do not give iron shots to piglets. Piglets voraciously eat the dirt, which supplies them with all the iron they need. Alternately, a shovelful of dirt (about a gallon or so) can be brought to the pen for the piglets to nibble on. Remember that iron is a trace mineral, and only minute amounts are needed to maintain proper blood cell formation. If given proper access, a piglet will eat what it needs for a balanced diet.

Feeding Lactating Sows

The goal in rations for a lactating sow is to provide enough nutrients for the sow to produce optimum milk outputs without depleting her vital nutrients and diminishing her physical condition. A lactating sow's daily nutrient load requirement is three times higher than that of a gestating sow.

When provided with the proper nutrition, a sow produces ample milk for her offspring. Piglets grow rapidly on the sow's milk.

Feeding three times the normal ration, however, is neither practical nor healthy. Therefore, a specific ration must be provided for lactating sows. Premixed mineral and vitamin supplements made especially for lactating sows can be added to feed for this purpose. Corn and soybean meal are also concentrated sources of energy and protein. High-fiber or high moisture-content feeds such as alfalfa hay, beet pulp, oats, or wheat add bulk to the diet but may dilute the nutrient intake. When fed by volume or weight, these high-fiber feeds should be fed as free-choice supplements to the lactating sow, not as a replacement.

An old-fashioned method of feeding the lactating sow is to provide the traditional corn-soybean ration with the appropriate supplements and bulky ingredients in a separate, free-choice feeder. Increasing feedings to three times per day encourages the sow to ingest the proper amount of nutrients.

Lactating sows need plenty of fresh, clean water as well. If a sow starts to dehydrate, her milk production suffers immediately. If the dehydration continues, the milk production will stop and, in most cases, cannot be revitalized. Be sure to provide water in a container that is piglet proof or sufficiently high to keep piglets from getting in and drowning.

Iron Shots for Piglets

Piglets are born with a low reserve of iron, and the sow's milk will not provide this needed nutrient. As mentioned earlier, if your pigs are raised on dirt and the piglets have access to the dirt, they will eat the dirt to gain the needed iron. Alternatively, a clod or shovel full of dirt can be brought to the sow's pen a few times per week.

Piglets raised on dirt generally do not need iron shots. Iron can be obtained from the dirt as the piglets eat and root around in it.

Advice from the Farm

Care of Piglets

Our experts offer valuable information about piglet care.

Grafting Piglets

"When several litters are born at the same time, the piglets can be grafted onto other sows to even out the litter sizes and the size of the pigs in each group. This will allow the runts to get their fair share of milk and ensure that the sows can feed all they are nursing."

—Robert Rassmussen, Rassmussen Swine Farm

Importance of Milk

"The first week is most critical; piglets need to get enough milk to get on their way. Sows that get up and down a lot smoosh pigs: let them farrow in a clean pasture. Let the pigs behave naturally, and let the mother raise them. Castrate at one day old; it's the best time. If the pigs are only a few days old and are not getting enough milk, you can put them under another sow with pretty good success. Extremely large litters can be shuffled to another sow with a small litter to even them out and give everyone a better chance."

—Josh Wendland, Wendland Farms

Piglet Pecking Order

"Piglets quickly establish a pecking order or teat order. Distributing piglets of equal size on different mothers will offer the smaller piglets a better chance of equal milk."

—Bret Kortie, Maveric Heritage Ranch Co.

Pigs in indoor pens or commercial facilities have to be provided the needed iron. Check with your veterinarian or local feed store for recommendations for your area. Read the label of the iron supplement carefully to be sure you understand the proper dose.

The easiest method of administration is by injection, and each piglet should be treated within the first three days of life. Have syringes and needles on hand for the task. Injecting a piglet in the ham is an easy, one-person job. Hold the piglet upside down by the rear leg that is to be injected. With the thumb of that hand, push the skin aside from the injection site in the fleshy part of the ham; inject the iron, remove the needle, and release the skin.

Watch for signs of anemia in your piglets. Symptoms include pale skin, rapid breathing, sloppy diarrhea, and weakness. If you see these symptoms, administer iron shots immediately, or contact your vet to administer the shot.

Identification

You will most likely want to establish a method of identifying each piglet, its parents, and its date of birth. There are a number of ways to do this, including ear notching, tagging, and tattooing.

Ear Notching

There are many systems of ear notching. If you raise purebred pigs, check with your pig's registry to see what system they require. In most systems, a particular right-ear notch indicates the litter number that the pig came from, and a particular left-ear notch indicates the pig's number in the litter—for example, 2 of 9. Ask your local 4-H club or county extension offices if they have booklets with notching instructions, or have an experienced pig person instruct you in the proper way to notch piglets.

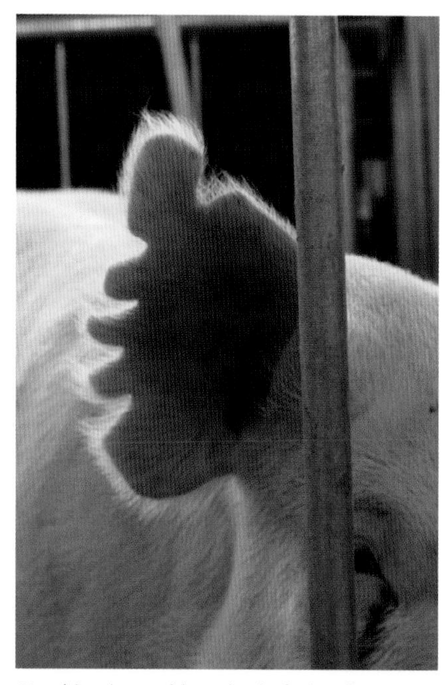

Notching is an old method of identifying pigs. Today, some consider it an optional procedure or even a maiming practice. A properly notched ear, as seen here, can serve as a ready ID in most cases.

A notcher is a convenient tool for this task, although small scissors or a razor blade could work. Remember to notch only about 1/4 inch, as the notch will get larger as the pig grows. If you are clipping needle teeth, you should also ear notch at the same time. These procedures should be done within the first 24 hours of life.

Tagging

Tagging is another method of identifying pigs. If the piglets are not mixed with other litters, you can wait to tag the piglets until just before weaning or mixing with others. Piglets can be tagged early, but a very small tag will have to be used. This may be inconvenient later on, as reading a small, dirty tag without catching the pig may be impossible. However, large tags used on small piglets will cause them to walk sideways or in circles until they get used to the tags. Tagging the piglets early may also upset the sow, and she may refuse to feed them or may even attack them.

There are many systems of tagging. Tags may be color coded, consecutively numbered, or differentiated by size or shape. Being able to look at a pen full of pigs and readily identify littermates by their colored tag makes color coding a popular choice. Regardless of the method you choose, tagging all the males in the right ear and all the females

Identification of individual pigs is important, especially in purebred herds. By tagging, notching, or using other identification methods, a herdsman can readily identify which piglets belong to which sow, what sex the piglets are, and so on.

in the left ear will make sorting and counting much easier.

Tattooing and Branding

Tattooing is usually reserved for large producers. People who already own a tattooer (used for goats, sheep, and cattle) may find that tattooing is the best method for them. White pigs obviously tattoo the best. Finding and reading a tattoo on a black hog may be next to impossible, and you will likely have to shave the pig's ear to read the ID.

A freeze brand is a newer type of pig identification. A branding iron, usually made up of numbers or letters, is "frozen" with liquid nitrogen or dry ice and alcohol. It is then placed on the skin for approximately 30 seconds. The brand will swell, eventually scab over, and then grow back hair void of pigment. The white brand mark will be readily identifiable. Obviously, this system will not be very effective on white pigs. It also requires more expensive equipment than does tagging. It is, however, a permanent identification that cannot be torn off. It is most popular in large herds that retain a high percentage of breeding animals long term.

Optional Procedures

Some optional procedures are performed on piglets in an effort to increase weight gain, prevent poor meat quality, and prevent injuries to the piglets or littermates. Procedures such as clipping needle teeth, docking tails, and castrating are performed at the discretion of the pig owner. If you decide that these procedures are appropriate for your herd, seek the advice and instruction of an experienced swine person. You may have the opportunity to visit another farm and learn these practices first hand. Otherwise, a veterinarian can assist you on your own farm. Try to perform the procedures together, as indicated below, to reduce the stress caused to the piglets.

Cutting Needle Teeth

Newborn piglets have eight needle teeth: two on each side of the top and bottom. These teeth are not permanent but remain in the pig's mouth until about six months of age, when the permanent adult teeth grow in. Needle teeth (sometimes referred to as wolf teeth) are very sharp and can be hazardous. These teeth may cause serious injury to the sow's teats by slashing or cutting, resulting in her unwillingness to nurse the piglets. Needle teeth can also cause injury to the other piglets as they spar with each other and jockey for space to nurse. Cuts can easily become infected or cause serious problems. Piglets that are runts may benefit from retaining their needle teeth, as this may give them some added advantage over the larger piglets in getting to nurse properly.

Gilts (first-time mothers) are most affected by needle teeth, as they typically have sensitive teats and are not used to the sensation of nursing. The teats on older sows become calloused or tougher, making cutting less likely.

Needle teeth can be cut to reduce the damage they can cause, but the procedure is not mandatory. Many swineherds with smaller herds do not cut needle teeth, with no adverse effects. Many sows and gilts nurse their piglets with ease, regardless of the needle teeth. After all, pigs in the wild have dealt with needle teeth for thousands of years, and piglets continue to survive into adulthood.

Piglets raised in large-scale operations are often castrated when they are one to three days old. By castrating all boar piglets, segregation of the sexes is not required during the growing and finishing phases.

The Castration Debate

Prior to your piglets' birth, you should have given consideration to practices by which you will raise the piglets. The proper methods and care of piglets is a highly debatable subject among pig breeders. The methods used are eventually determined by personal preference. Care is taken here to represent both sides of the argument in an effort to allow readers to make an educated choice regarding the care of their own piglets.

According to the Farm Animal Welfare Council, "mutilation practices" such as tail docking, nose ringing, clipping needle teeth, and castration of boars are inhumane and undesirable. The council does, however, recognize the need for such practices under management conditions in which the practice may help avoid worse problems (such as aggression, tail biting, or teat slashing by piglets).

All these practices do serve a purpose and have been utilized for many years. It is important for the breeder to know the whys of each practice to make a decision about enacting them.

Most producers castrate boar piglets before fourteen days of age. A physical change in boars when they reach sexual maturity can produce a hormone that gives an off flavor to the meat, known as meat taint. Castrating the males early means the hormone is never produced, thereby eliminating this contamination. Additionally, many people believe that castrated boars are less aggressive. Furthermore, castrated boar piglets can be housed with females without the threat of breeding. This may make penning more manageable, especially for a producer with limited space.

Conversely, castrating boar piglets is stressful to the pigs as well as to the breeders. If you are not an experienced handler, you will need the assistance of a veterinarian (or an experienced pig raiser), adding cost to your production. Many modern hog raisers believe that castrating actually makes a fattier pig. The castrated males (barrows) can become lazy or lethargic, eating but not exercising, and thus putting on less muscle and more fat. Cutting the food ration to compensate for the extra fat production will result in a longer amount of time to raise your pig to butcher weight.

Make an informed decision about your own practices by getting as much information as you can on the subject.

If you decide to cut needle teeth, the following method is easiest and most efficient. Needle teeth need to be cut within the first twenty-four hours of the piglet's life. If you trim the teeth within six hours of birth, the piglets won't screech too much, and you are less likely to raise the sow's ire.

Nail clippers for humans or cats, small side cutters (pliers with an angled blade), or cuticle shears are the best tools for this task. Holding the piglet

with its back against your chest, place your thumb to the far back of the piglet's mouth to pry it open. Snip the needle teeth even with the other teeth. Take care not to cut the gums or tongue, and do not cut the teeth off at the gum line. Mark each piglet with a grease stick so that you know the piglet has been clipped.

Docking

Docking (shortening the tail) is typically performed in facilities that raise large numbers of animals in indoor, crowded spaces. Overcrowded pigs tend toward aggression, and tail biting is one of the ways in which pigs act out on each other. On a farm that has plenty of room for growing piglets, and where the space at the feed trough is sufficient to service all at the same time, docking may be an unnecessary procedure.

If you choose to dock, remove only the bottom third of the tail, the part thought to be less sensitive. Snip the tail with wire cutters or similar tool. The wound should be treated with a disinfectant. No anesthesia is used in commercial facilities. If you dock, do it within the first twenty-four hours of the piglet's life.

Castrating

Castration is a traumatic experience for boar piglets. It is recommended that it be done while they are still suckling, preferably a week or so before weaning. If you wean your piglets early (before eight weeks), you may have to wait for the testicles to be sufficiently large to perform the procedure. Commercial piggeries typically castrate at three days of age, but these are experienced handlers who have castrated pigs many times before. Ask a vet or other experi-

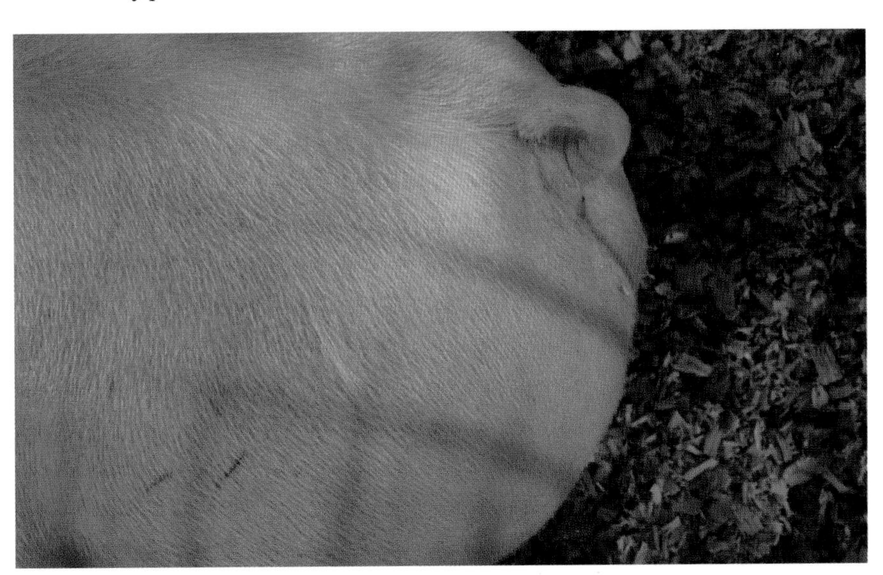

Docking tails can prevent tail biting in some herds. Docked tails should have only the lower one-third removed. Most show pigs have docked tails.

enced person to show you how to perform this procedure.

Assemble your castration tools ahead of time. A new razor blade or scalpel is required, as is disinfectant for the wound. (Topical treatments and bandages are not necessary.) Castration takes two people or the use of a sling or castration box. A sling is a V-shaped or curved cradle in which you lay a piglet on his back, then use a rod or brace to hold the back legs forward, exposing the genital area. The brace is placed just behind the hock. Some cradles come equipped with a storage shelf and hanging device so it can be hung on the wall of the pen. A castration box is a unit in which the piglet is suspended from the back legs by placing the legs into two straps and tightening. The piglet relaxes once hanging, and the straps keep the legs far enough apart to allow you to perform the castration. Either unit is also handy for giving shots, clipping teeth, or any other chore that requires more than two hands.

If you are using two people to castrate, place the piglet on his back in the lap of the holder. Have the holder keep the back legs held forward (toward the holder). This secures the piglet and presses the testicles tightly against the scrotum.

Swab the area with a disinfectant. On one side, press the testicle against the skin between the thumb and forefinger. Make one cut, approximately 1/2 to 1 inch long, to expose the testicle. Make sure you cut low on the scrotum, which will actually appear high on the overturned pig, to ensure proper drainage of the wound.

As soon as the cut is made, gently pull the testicle out and away from the body. It will be attached to two cords; one is the sperm duct, the other is the blood vessel. Continue pulling the testicle out until resistance causes the artery to snap off. This may sound cruel or disgusting, but the blood vessel automatically shrinks back into the body, stopping the blood flow. Much more blood will be seen if the artery is actually cut.

Repeat the procedure on the other side. No stitches are needed. The wounds can be treated with disinfectant or fly spray if desired, but this treatment is not necessary unless the piglets are kept in unclean facilities. Do not dress the wounds in heavy creams or petroleum based products, as they will inhibit proper drainage and may induce infection.

Place the pig in a clean, dry, freshly bedded pen. He will likely run to his littermates for consolation or try to nurse. As long as the castrated piglets remain active, alert, and interested in food, you can rest assured that they are not in any distress. Some swelling may appear, but it will go away within a few days.

If excessive swelling, bleeding, or pus is evident, contact your veterinarian for proper treatment.

Food, Water, and Amenities for Piglets

Providing the best possible living quarters, feed, and water source for your

Growing piglets should be housed with ample space for reclining and roaming. Trough or feeder space should be adequate to allow all piglets to eat at once in hand-feeding systems, or at least half of the piglets to eat at the same time in self-feeding systems.

piglets are the key to weaning the maximum number of piglets per litter.

Creep Feed and Water

Piglets should be provided creep or starter feed, formulated specifically for very small pigs, within the first ten days of life. Starter feed is milk based, as opposed to a soy-based feed. Milk-based feed is highly palatable and easily digested by piglets. Milk solids provide the needed protein (amino acids) for sufficient piglet growth and energy. Plant-based proteins are not easily digested before the piglets are about two months of age, as their intestinal flora has not developed sufficiently to utilize the plant protein.

Creep feed should be placed in an area that only the piglets have access to, such as under a rail or near the heat source. Most will not show interest in the feed right away. Provide only a small amount at first, and refresh or stir it every time you do chores. Something new will pique the piglets' interest, and they will eventually start to nibble and eat.

Water should be provided specifically for the piglets in very shallow pans or in a nipple drinker. Piglets usually start to eat and drink at about two weeks, although some take much longer.

Your piglets must be eating well for a week or two before weaning to ensure that they will not lose weight once they stop nursing. Additional milk can be provided at weaning, but is usually not necessary unless a piglet has stopped eating altogether.

Bedding

When the piglets are nursing, it is better to add bedding on top of the existing bedding in their pen than attempt to replace old bedding. If the bedding is terribly wet and soiled, some removal might be in order, but proceed with caution. Disturbing the sow while she is still nursing may cause her to become agitated and aggressive toward you or the piglets. Sows may become so disturbed by someone cleaning their pens that they savage (also known as cannibalize) their own piglets. Clearly, it is best to let them have more bedding than to create such a situation.

Heat Source

Piglets do best when they are maintained at a temperature of about 70 degrees Fahrenheit. This is not practical for most people, especially with winter farrowing. Do your best to provide plenty of bedding, minimize drafts, and provide a steady heat source. Heat lamps are effective for even the coldest of barns. Set the lamp low enough to provide heat for the piglets but high enough that the sow cannot chew on it. If at all possible, create an area that the sow does not have access to, about three to four feet from her resting place. Place the heat lamp in the restricted area. The heat and light will lure the piglets in. The piglets will rely on the heat lamp for warmth instead of trying to lie tightly against the sow. This will help to minimize the number of piglets that get crushed or stepped on when the sow moves.

Piglets readily take to a starter ration that is milk-based. The flavor encourages them to eat and prevents weight loss at weaning time.

Heat mats are also an effective means of keeping the piglets warm. Heat mats are rigid plastic mats with an internal heating element. The piglets lie on the mats for warmth. No additional bedding is needed on the mat. Heat mats can be expensive, so be sure that you place them well out of reach of the sow to prevent her from chewing and destroying them. Likewise, power cords for heat sources, lights, fans, and so forth must be kept well away from both sows and piglets to prevent chewing, electrocution, and the possibility of starting a fire.

Protection

You will be amazed by how quickly your piglets grow. Seemingly in no time, they will venture out to explore their surroundings. Piglets are master escape artists, but fortunately they never stray far from the milk tap and return frequently to make sure the sow is still there. Do not allow your piglets to roam into other adult pig pens. Pigs are known cannibals, and a bite-size morsel may be too difficult for an adult pig to resist. You can use panels specifically designed for piglets to keep them with the sows. These panels have small openings at the bottom, about 2 inches by 4 inches, that piglets cannot climb through; a typical cattle panel will not suffice. Boards along the bottom of the pen will also keep the piglets in. They can climb over partitions up to 18 inches high, so be sure to have your boards adequately placed to prevent the piglets from climbing over.

Weaning

When it is time to wean the piglets, it is best to remove the sow to her new pen first, usually to be returned to the boar. If the piglets are to be moved to new pens also, do this after the sow has been taken out of earshot. You will have a much more difficult time removing piglets from a pen if the sow is still in it, and she may get aggressive toward you.

In addition to enduring the stress of being taken from the sow (weaning), making the transition from milk to a dry feed ration is very challenging to young pigs. Your pigs should be eating well from the creep feeder prior to weaning to make this transition easier. Additionally, the amount of liquid provided by sows' milk needs to be replaced with water. Piglets drink a large amount of water for their size. Ample sources of water, sufficient to allow all piglets access as needed, are an absolute must.

The main goal in feeding starter pigs is to provide ample nutrition for rapid growth. Fresh and dried milk products contain proteins that are highly palatable and digestible to young pigs. Soybean meal may cause allergy symptoms in

All in the Timing

Try to wean your sows on a Thursday. The sow will usually come into heat on the following Monday. Placing the boar with the sow on Monday will make your due date calculation much easier. Your sow should farrow midweek on her due date.

very young pigs, such as diarrhea and weight loss. Only a very small portion of the starter feed should contain soybean meal (less than 20 percent). After about two weeks on the starter ration, the soybean meal portion may be increased.

Phase feeding is practiced by many commercial pig facilities. Phase feeding involves feeding several diet formulations in quick succession, tailored to the growing piglet's nutritional needs. This system is expensive and impractical for the small farmer raising only a litter or two per year unless commercial rations can be purchased in small amounts from a local feed mill.

While living with the sow, the piglets were likely already eating the sow's ration. The addition of milk products will help balance the piglet's nutritional needs. Unless you wean your piglets at a very early age (less than twenty-one days old), you will be able to feed them the sow's ration with no ill effects. If you practice early weaning, it is best to buy a ration specifically designed for early-weaning pigs.

There may be a time when you must wean piglets early, such as when a sow is not producing milk or has cannibalized her piglets. You can properly provide for your piglets by feeding about one cup of whole milk per piglet in addition to the creep feed, or mixed with the creep feed as a mash. Piglets relish a warm, juicy mash. Do not feed more than they can clean up between feedings, and do not allow the mash to become sour or moldy. If the mash is heavily covered in flies or has a sour smell, discard it immediately.

Conclusion

As you can see, rearing piglets to the weaning stage has many phases and tasks. By practicing these fairly simple methods, you will have great success in raising piglets. Seeing a group of healthy, fast growing piglets is one of the true pleasures of raising pigs.

CHAPTER EIGHT

Pork Processing and Butchering

Delicious home-cured meats are the highlight of any hobby farm or homestead that produces its own pork. There are substantial secondary benefits as well: you can be confident that your meat is wholesome, since you know how it was raised, and you will have taken one more step toward a self-sufficient lifestyle.

Pork is highly nutritious. According to the Virginia Cooperative Extension bulletin #458-921, *How to Stretch Your Pork Dollar*, a mere 3.5-ounce serving provides over half the protein; 74 to 103 percent of the thiamin; 18 to 37 percent of vitamins B6, B12, niacin, riboflavin, and phosphorous; and 19 to 35 percent of the iron an average adult needs daily. Furthermore, a 3.5-ounce serving from the loin, roast, chop, ham, or tenderloin contains only about 250 calories.

You may be wondering just what you will end up with after raising and butchering your own pig. A pig is definitely not all pork chops! A typical butcher hog, at 250 pounds live weight, will have a carcass weight (also known as hang or hanging weight) of about 180 pounds, or 72 percent of live weight. This weight is composed of:

- Retail cuts: 145 lbs (58 percent of live weight)
- Fat, bones, and skin: 35 lbs (14 percent of live weight)

Carcasses shrink during the hanging process, as they lose moisture and dry slightly. The carcass weight includes the leaf lard and kidneys, which will not be reflected in the retail cuts of meat.

Retail cuts (also known as freezer weight) would be approximately:

- Ham: 44 lbs
- Loin chops: 36 lbs
- Cured or fresh picnic roasts: 12 lbs
- Boston butt roasts or slices: 13 lbs
- Bacon: 28 lbs
- Spareribs: 7 lbs

Delicious products can be made from pork, such as these prosciutto hams. Prosciuttos are salted and air dried for months and sometimes aged for up to one year.

Additional meat is obtained by saving the feet, tail, neck bones, organs, and trim (used for sausage meat), approximately 25 more pounds.

A look at the following chart may help you visualize exactly what meat cuts come from which part of the hog. Study the chart carefully before you butcher or have your hog butchered. You will be able to determine what cuts of meat you want from the carcass so you can tell your butcher.

You need to decide whether to take your pigs to a butcher or butcher them yourself. The decision depends on various factors, including your ability to process the hog, your personal feelings about killing an animal you have raised, your storage space, and the availability of the tools to get the job done.

Scheduling the Butchering

Fall is the traditional time of year for butchering on the farm. This coincides with the natural harvest cycle, and the cooler temperatures help meat chill quickly and thoroughly. If your meat will be processed by a butcher shop, you can have your hog slaughtered at any time of the year.

Several weeks before you intend to slaughter, evaluate your pig for readiness. Pigs fed to a weight of 225 pounds are ideal. Pigs raised over 225 pounds tend to put on more fat than muscle. Feeding for additional weeks to gain more fat is a waste of money and effort unless lard is your primary goal. A pig gains approximately 1.8 pounds per day from the size of 150 pounds to 225 pounds. At this rate, a hog weighing 190

Cured and aged salamis, hams, sausages, and more are delicious ways to preserve pork for long-term storage as well as to add tasty variety to your menus.

Meat Cuts

Boston Butt	Blade Boston roast, boneless blade roast, blade steak, shoulder roll, cubed steak, shoulder chops, pork cubes, neck bones, fat back, lard
Foot	Pickled or fresh pigs feet
Ham	Boneless leg, boiled ham, smoked ham, canned ham, ham slices, smoked ham: rump or shank
Hock	Smoke for soup bones, pickle, fresh hocks
Jowl	Smoke jowl or cheek, bacon
Loin	Blade, sirloin, rib, loin, butterfly or top loin chops, country-style ribs, back ribs, smoked loin chops, Canadian-style bacon, boneless top loin roast, tenderloin, blade loin, center loin, sir-loin, crown roast
Picnic	Smoked arm picnic, arm roast, ground pork, fresh arm picnic, arm steak, sausage
Side	Spare ribs, slab bacon, salt pork, sliced bacon, lard
Trim	Sausage, pork cubes, ground pork

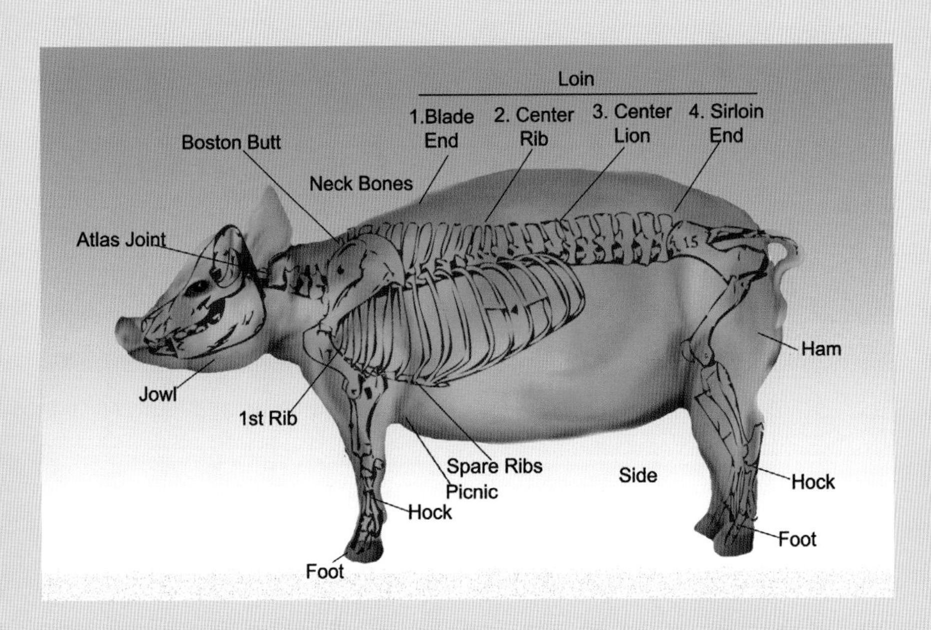

Advice from the Farm

Processing Your Pigs

Experienced meat producers give their advice on selecting a processor for your pigs.

Get Samples

"Sample a variety of cured products from your potential butcher. If you don't like the flavor of his products, your customers probably won't like it either. Customers always relate the quality and flavor of the meat back to the farm. If your butcher isn't good, they will think it is the fault of the meat. Many small towns have custom butchers, which means you can have the meat processed for yourself but cannot sell it. If you want to sell the meat, you'll have to look for a plant that is federal or at least state inspected.

If you can load the pigs the evening before, and they are not too crowded on the trailer, they will be more calm when it is time to move them. Provide lots of water, and park the trailer in the shade. Once the trailer starts moving, the pigs generally just lie down and aren't bothered by the movement. Don't holler at them or poke at them to get them moving. Take it slow and they will eventually load or unload in the direction you want them to go."

—Bret Kortie, Maveric Heritage Ranch Co.

Interview the Butcher

"Sanitation in the processing facility is the number one thing to evaluate. Look for a federal- or state-inspected plant. Visit the site and look for things like cleanliness; for example, are the tables clean, is there debris on the floor, does the help have good hygiene? When looking for a butcher, inquire first how many years he has been in business, what type of experience and training he has, and any specialties he may have. An experienced butcher will give you a higher quality and better-looking product. Your butcher should be someone who is easy to work with; you have to get along with the person so that they understand what you want in your processing.

When ready to butcher, make sure your pig is ready. By this I mean that the hams are well rounded and briskets are filled in, loins are round, and a good fat cover is on your pig. I recommend taking the pigs off feed at least 12 hours prior to butcher. But, pigs must have a lot of water; this is so important. Give them all they will drink. You don't want them to dehydrate before slaughter. The meat gets off color, fibers are drier, and your quality goes down drastically. You must keep moisture in meat for a good end product.

When transporting the pigs to the butcher, don't wind the pigs up, chase them around, or use a shocker on them. Keep them calm and try not to stress them."

—Doug Klarenbeek, Hudson Meats

Pigs of uniform size and market weight are ideal candidates for processing. Spring-born pigs are usually ready for processing by late fall, an ideal time to cure meats.

pounds will take about twenty days to go from its current weight to the ideal weight of 225 pounds. Knowing this, you can determine your slaughter date within a few days.

The exact instructions on home butchering are beyond the scope of this publication. If you are interested in processing your own meat, refer to the Morton Salt Company publication *Home Meat Curing Made Easy.* This booklet, originally published in 1941 but updated many times through the years, is the standard to which all home processors should adhere. In it, you will find everything you need to know about the butchering and curing process, not only for pork but also for beef and lamb. The volume gives lists of equipment needed in addition to step-by-step instructions and photographs.

Additionally, *Butchering, Processing & Preservation of Meat* by Frank G. Ashbrook and *Sausage & Small Goods Production* by Frank Gerrard are excellent resources for the home processor and meat preservationist.

Selecting a Processing Facility

If you live in an area that has many processors, you can take your pick for

Estimating Weight

You may be wondering how to determine the weight of your pig if you do not have access to a scale. One way to approximate the weight is to measure the heart girth (the distance around the pig in the "armpit" area) and the length, measured from the point midway between the ears to the base of the tail, and then calculating as follows:

Heart girth x heart girth x length divided by 400 = estimated weight

the one to process your pork. When evaluating a butcher, look for one that is licensed under the appropriate rating, such as Custom, State Inspected, USDA Inspected, or Organic, to accommodate your intended purpose for the meat. Speak with prospective butchers to obtain their ratings. Remember that if you intend to sell your pork, you will need to have it processed through an appropriately rated butcher.

All butchers have their own methods and recipes for curing ham and bacon and for making sausage and other cured meats. The best way to identify the butcher who processes the meats to your taste is to sample his product. Most small shops have meat products available for sale. If not, ask for samples. By sampling the items you know you want to have, you can determine who makes the meats most pleasing to you. Sample ham, bacon, and sausage, in particular, as these will be your primary cuts of cured meats. This is all a matter of preference, but by sampling, you can compare and determine who uses too much salt, makes the ham too smoky or not smoky enough, makes sausage with just the right amount of spice, and so on. This step is imperative if you will sell your pork products. Your product reflects on you, not the butcher. If your processor does not make a product that is pleasing to your customers, you will lose future sales. Select a butcher who makes a product you will be proud to call your own.

When selecting a butcher, inquire as to the packaging methods used. If you are raising your pork strictly for home use, you may be perfectly happy with

Once hams have been properly cured and aged, they are ready for purchase. Specialty hams such as those shown here can command up to $20 per pound.

Butchering Shops

Many towns and most larger cities have butchering shops that serve the small producer. Such shops are officially rated as either Custom, State Inspected, or USDA Inspected. Custom shops can butcher and process only animals that are brought in and will be consumed by the same person. The meat cannot be sold. It is packaged and labeled Not for Sale. Custom shops usually process wild game as well as farm-raised animals.

Rules of state-inspected facilities vary, depending on the applicable state laws. In most states, processed meat can be individually packaged and sold within the state, as it is inspected by a state regulatory agent.

USDA-inspected facilities operate under the USDA's jurisdiction, and all animals are inspected prior to and after slaughter by a USDA-certified inspector. Unless prohibited by your state, USDA-inspected meat can be sold anywhere in the United States. Processing meat at an inspected facility will cost more, as there is a per-animal inspection fee plus additional charges for any packaging or labeling required as the result of the inspection.

Some processing facilities are making the move toward becoming certified as organic. Organic-certified facilities can process animals that have been raised by certified organic farms as well as nonorganically raised meat, but the two products must be processed completely separately, and all equipment must be cleaned and sterilized between processing nonorganic and organic animals. By processing only organic meat, the facility removes the possibility of any meat that has been fed hormones, antibiotics, or animal-based products from coming in contact with the organically raised meats. There are only a handful of organic-certified butchers in the United States. Farmers should expect to pay a premium for processing at these facilities, as the organic certification adds to the costs of the butchering facility. Organic certification agencies can provide a list of butchers who have been certified.

meat packaged in white butcher paper. If you sell the meat, you may prefer to have the meat sealed in cryovac packages. Meat is vacuum sealed in clear plastic that is more sturdy when being moved about in the freezer and also allows a person to see the meat and identify it. Many consumers prefer to see what is in their package, so this should be a consideration when processing. Vacuum-sealed packages cost more to produce, so expect to pay a bit more for this type of packaging. Look at sample packaging if possible, and evaluate it for eye appeal and the ease

Many regional and ethnic-type sausage recipes are available. Experimenting with recipes helps you determine which you prefer and which your customers will be interested in buying.

with which the meat can be identified. Look at it from your customer's point of view. Does the package look good, clean, and appetizing? Then it will probably be pleasing to your customers.

Some butchering facilities will allow the customer to be on site when they are processing an animal. Ask if you can be present when a pig is processed, so you can watch the method and understand how a pig goes from farm to freezer. If you are uncomfortable with what you see or feel that the facility is lacking in some way (such as cleanliness or staffing), look elsewhere.

Make arrangements with a butcher for processing your pig at least one month in advance of when the pig is ready. Although your processing will take place at the facility, you should become familiar with the various cuts of meat to give your butcher specific instructions on what you want in freezer cuts. Most butchers will assume that you want them to cure your hams, bacon, and sausage. If you intend to make your own cured pork products at home with your own

Did You Know?

Crubeens is the English word for the hind feet of a pig. Served pickled, boiled, roasted, hot, or cold, crubeens are prime pub fare. According to food writer and historian James Beard, they have a delicious flavor, tender texture, and a pleasingly gelatinous quality.

recipes, notify the butcher that you want the fresh (green) hams and bacon for home curing, and be ready to pick them up as soon as the carcass is chilled and ready to be cut or cured. Keep in mind that your home-cured meats will be for your consumption only and not saleable in most areas.

About two days prior to slaughtering, move the intended butcher pig to the holding pen to ready it for transport. You can even load the pig onto the transport vehicle and allow it to settle in. Gently handled pigs that are not stressed, overheated, agitated, or frightened will produce much finer and mellower tasting meat. The pig can be washed down if necessary at this time. Remove feed from the pig at least twenty-four hours before slaughter, but provide ample amounts of water. Feed withdrawal makes removing the viscera easier and minimizes the chance of contamination from the stomach contents, as the gut will not be full of fermenting food.

When transporting your pig to the processing facility, take care to prevent the pig from getting overheated. Most pigs lie down when being transported. Unless they are excessively crowded or being transported with pigs they are not familiar with, the pigs will remain calm. Do not transport two or more boars that have not been housed together, as they will engage in fighting, sweating, and other dominating behaviors. Give the pig time to walk off the trailer or truck at his own pace, without being chased or hurried. If you can deliver the pig the night before slaughter or very early the morning of, the pig will have time to settle down.

Conclusion

As you can see, fully researching potential butchering facilities and spending time getting your pigs prepared for butchering will greatly improve your products. By producing and marketing the absolute best products possible, you ensure having repeat customers and attracting new ones.

CHAPTER NINE

Marketing Your Pigs

According to the USDA Economic Research Service, the United States is the third largest producer and second largest consumer, exporter, and importer of pork products in the world. American pork consumption per capita is highest in the Midwest (58 pounds), followed by the South (52 pounds), the Northeast (51 pounds), and the West (42 pounds). These averages vary within different ethnic and age groups.

Consumers are becoming more educated on their purchases, and recent trends indicate that people are more inclined to purchase products from locally grown sources. Consumers are also taking an interest in the source of their food and seeking to connect with the farmers who are raising it. Being able to trace a food to its source is a key selling point for many hobby farmers and small producers.

Marketing Pigs as Meat

The first order of business in selling meat in the United States is to research and learn all the applicable laws for your state. Most states do not allow the sale of meat that has been processed in another state unless it was handled through a USDA-inspected facility. If you intend to sell all your meat within your state, you may be able to use a state-inspected facility for processing. If selling an animal on the hoof, a custom butcher shop is within the law. Selling meat by the cut or package is usually governed by USDA standards and must be processed as such.

Selling a pig on the hoof is the easiest method of marketing your hogs. This means the pig is sold live and delivered to the butcher or picked up by the customer. Pigs sold in this manner become the property of the buyer immediately, and

What Is Organic?

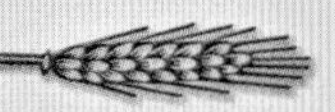

According to the USDA National Organic Program, foods that are labeled Organic must conform to the following standards:

"Organic food is produced by farmers who emphasize the use of renewable resources and the conservation of soil and water to enhance environmental quality for future generations. Organic meat, poultry, eggs, and dairy products come from animals that are given no antibiotics or growth hormones. Organic food is produced without using most conventional pesticides; fertilizers made with synthetic ingredients or sewage sludge; bioengineering; or ionizing radiation. Before a product can be labeled 'organic,' a Government-approved certifier inspects the farm where the food is grown to make sure the farmer is following all the rules necessary to meet USDA organic standards. Companies that handle or process organic food before it gets to your local supermarket or restaurant must be certified, too."

all processing and transport are now the new owner's responsibility. Offering to deliver the animal to the local butcher may seal a deal for you. A lot of people living in town do not want to be introduced to their future food, nor do they have the vehicle for transport.

Assisting potential buyers in choosing the cuts of meat is another helpful service. First-time buyers are often uneducated about the meat that comes from a pig, thinking it all chops, bacon, and ham. Letting them know they have options may be of great benefit to them and will also establish a relationship of trust as you pass along your knowledge.

Many people also do not have any idea of the cuts and processes involved in butchering a pig. Your knowledge will be invaluable to them. By being able to intelligently discuss how their pork can be cut, processed, and packaged, you will ensure that your customers will be happy with their purchase.

Talk to the butcher you will be recommending to your customers. Get a list of any specialty products they produce from pork, such as sausages, hams, crown

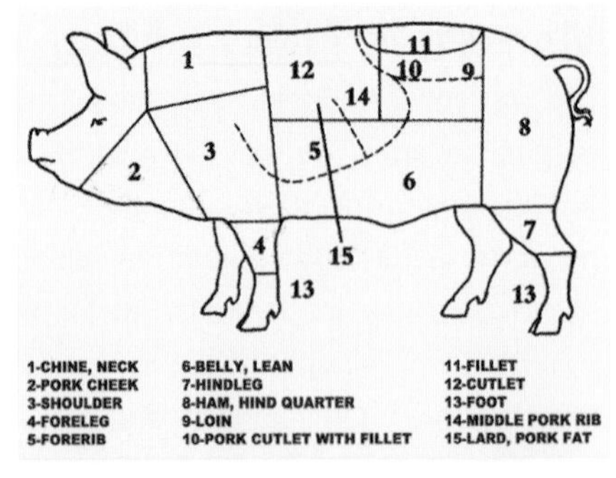

Most people think of chops, ham, and bacon when they think of pork. But a large variety of specialty meats can be produced as well.

Pigs raised under natural or organic systems are becoming more popular with consumers. Organic pork can garner up to ten times the price of commercially raised products.

roasts, and smoked meats. Be willing to sample many of his products so you will be able to talk about them with your customers. If your butcher uses a standard cutting form, get a copy of that also. Standard cutting forms are lists of the usual or most popular cuts of pork, the amount you want in your packages, and the size of the portions (that is, thickness of the chops). This will help you discuss the usual cuts of meat and make it easier for customers to decide which cuts they want from the pig they have purchased.

Remember that people will assume that anything that goes wrong with their pork purchase will be attributable to your farm, as you are the selling agent. Do your best to screen and test the butcher before recommending him.

Marketing Organic Pork

If your desire is to market pork labeled Organic, you need to be certified by a recognized organic certification organization and need to label your product as such. The certification process involves a lengthy application that will inquire about your production methods; sustainable practices on your farm; feed sources used; any fertilizers, chemicals, or supplements used; and your general husbandry methods. You will be required to comply with all standards as established by the USDA National Organic Program (NOP) as well as any additional requirements by the certification board. The best place to start your certification process is to go to the NOP Web site and review all the rules

Advice from the Farm

Marketing Your Meat

Experienced pork producers and marketing experts give their advice on marketing your pork.

Tell A Story

"When trying to sell your pigs, tell people the story about them. Highlight the benefits of buying your pigs and why your pigs are special. People love a good story. Tell them one."

—*Dr. Vincent Amanor-Boadu*

Know Your Goals

"Creating your pig production model takes time. Be willing to make short-term sacrifices for long-term goals. You may have to give up a few personal extras in order to get your facility producing, but it will be worth it."

—*Angela Peters, Madison, South Dakota*

Sell the Best

"Only the top 30 percent of animals should be sold for breeding stock; the rest should go for meat. Be prepared to stand behind the stock you sell. Sell only the best, the ones that will produce the best offspring. Selling whole butcher hogs is the best way to sell meat; just deliver it to the locker plant and the butcher takes it from there."

—*Josh Wendland, Wendland Farms*

Clean Meat Brings More Money

"The U. S. consumer is rapidly becoming aware of health issues associated with confinement hogs being fed growth hormones and given routine antibiotics. Being able to offer a healthy alternative, such as clean meat, is leading niche farmers to success. Advertise your meat as the wholesome product that it is, and you will attract discerning customers. Charge a premium price for this quality; clean meat has value, and you should be compensated for raising it."

—*Bret Kortie, Maveric Heritage Ranch Co.*

and regulations regarding certified organic products. They can be found at http://www.ams.usda.gov/nop/indexIE.htm.

According to the Organic Production and Handling Standards, "Animals for slaughter must be raised under organic management from the last third of gestation, or no later than the second day of life for poultry. Producers are required to feed livestock agricultural feed products that are 100 percent organic, but may also provide allowed vitamin and mineral supplements. Producers may convert an entire, distinct dairy herd to organic production by providing 80 percent organically produced feed for nine months, followed by three months of 100 percent organically produced feed. Organically raised animals may not be given hormones to promote growth, or antibiotics for any reason. Preventive management practices, including the use of vaccines, will be used to keep animals healthy. Producers are prohibited from withholding treatment from a sick or injured animal; however, animals treated with a prohibited medication may not be sold as organic. All organically raised animals must have access to the outdoors, including access to pasture for ruminants. They may be temporarily confined only for reasons of health, safety, the animal's stage of production, or to protect soil or water quality."

With the exception of farms producing certified organic feed, most small farms already comply with the bulk of the standards of organic certification. If you are able to purchase organically certified feed, you might consider applying for organic certification to boost the sales potential of your products.

The certification application process can take up to several months. Costs range from $500 to $3000, depending on your product. An on-site inspection of your farm is required, as are annual dues and a royalty fee on the number of products you produce. With this certification, you can label your product Certified Organic, either through the

Cured meats, such as pancetta, sausage, salami, and pepper-cured bacon, are all delicious ways to use pork. These specialty meats will be appealing to many gourmet chefs as well as to the lover of hearty sandwiches.

Specialty Product Ideas

Many delicious items can be made from pork. Consider one of these specialties to create your own niche market.

- Andouille (sausage)
- Boudin (sausage)
- Bratwurst (sausage)
- Canadian bacon (ham)
- Caul fat (the lacy layer of fat around the stomach)
- Chorizo (sausage)
- Coppa (Italian spicy ham)
- Country ham
- Galantines (finely ground meat, spices, and wine, pressed into a loaf and served thinly sliced, similar to meat loaf)
- Guanciale (unsmoked Italian bacon made from the jowls)
- Head cheese (gelatinous loaf made from the head and other scrap parts of the pig, pressed into a loaf, served as lunch meat)
- Honey ham
- Hot dogs
- Hungarian sausage
- Hunter sausage (similar to snack sticks or Slim Jim snacks)
- Italian sweet and hot sausage
- Jagerwurst (sausage)
- Kielbasa (sausage)
- Landjager (Bavarian semi-dried sausage, long keeping)
- Lard
- Lardons (ribbons of lard used to add moisture to drier cuts of meat)
- Mortadella (Italian bologna)
- Pancetta (Italian bacon, salt cured and dried)
- Pepper-cured bacon
- Pepperoni
- Pig ears
- Pork belly
- Prosciutto (mild Italian ham)
- Rillettes (French, meat slowly cooked in fat, then shredded and made in to a paste or spread)
- Salami
- Salt pork
- Seasoned pork chops
- Smoked ham
- Smoked hocks
- Summer sausage

certifying agency or the USDA. Some reimbursement for costs associated with the certification process may be available through your state or USDA office. Check with your certifying agency to see if your state or region qualifies for cost share reimbursement.

Once you are certified, you will also need to find a certified butcher to process your products. A list of certified butchers is available through NOP, as are the standards by which they must process your meats.

Selling Processed Pork Products

Not everyone has to join the niche-pork or specialty-marketing band to sell pork products. Selling standard products from your pigs is a perfectly legitimate way to offer pork of higher quality than feedlot-produced products.

First, evaluate what cuts and quantities you will sell. Plan to offer a limited number of options in products. For example, offer a pork package that includes chops, hams, bacon, roasts, and one or two types of sausage. This package simplifies relations with your butcher as well as with customers, as each pig will be cut and processed the same way. Again, check the regulations in your state for selling meat by the package.

No meat lends itself better to specialty products than does pork. Pigs have been used from nose to tail for centuries, and they have been creatively turned into delicacies in every nation that raises them.

Specialty products such as sausages and hams will likely need to be processed through a USDA butcher. All recipes for intended products must be reviewed and approved by a USDA inspector prior to marketing the product. Labeling is also a consideration when marketing specialty items. Labeling is also regulated by the government and needs approval before marketing.

To comply with the regulations of marketing under a brand or specialty label, seek the advice of a USDA-inspected butcher or the meat processing department of your state university.

Pointing out the qualities of your stock to potential buyers helps educate them about the finer points of your pigs and encourage a sale. Sell only the best you have to offer. Repeat customers and referrals are your best future sales opportunities.

Colleges offer many helpful brochures and classes relating to marketing meat products in your state. Usually, they offer meat-quality testing and will help with recipe development as well.

Testing the waters with your products can be done in several ways. The obvious first step is to give some of your product to family and friends for sampling. Pay attention to their feedback so you can improve your product. Marketing products is a very daunting task, which is another reason for limiting your initial offerings. That way, you can get a handle on all the requirements and any unforeseen problems.

Ethnic markets may offer a very good client base for your specialty products. Consider items such as salami, andouille, brats, or other ethnic specialties in your product line. The method of processing may be altered to accommodate certain ethnic markets. For example, Hispanic cooking includes the skin of the pork, the whole backbone (vertebrae), and neck bones. These items are usually discarded during processing, but they may offer a specialty market in your area.

If you raise purebred pigs or one of the endangered or heritage breeds, you should advertise your pigs as such to draw in specialty buyers. Each breed's

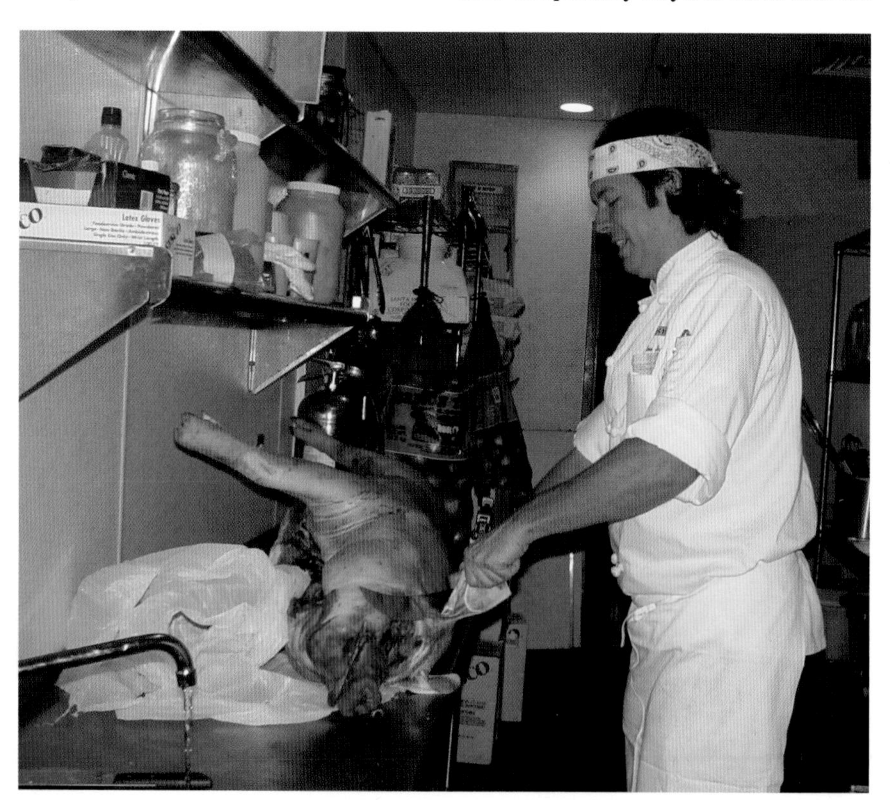

The whole hog can be used to make everything from chops to souse (head cheese). Cochon Restaurant in New Orleans offers delicious gourmet dishes using everything but the squeal.

Unique Pork Products

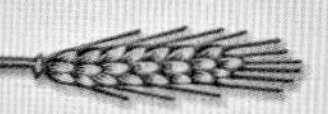

People who raise pigs for home use brag that they can use everything but the squeal! Indeed, there are a variety of products made from the pig that surely utilize every last bit.

Part of the Pig	Product
Bones	Buttons, sewing needles, pet food
Bones and skin	Pigskin garments, gloves, shoes, and footballs
Bristles/hair	Brushes for hair, teeth, shaving, and artists; insulation
Caul fat	Wrapping roasts and poultry
Fatty acids and glycerin	Insecticides, floor waxes, weed killers, water proofing agents, cement, rubber, crayons, cosmetics, chalk, antifreeze, plastics, putty, cellophane
Head	Cold cuts (lunch meat), souse
Knuckles, feet, ears	Gelatin, pickled products, pet food
Leaf or kidney fat	Lard
Liver	Terrine or French paté
Meat	Chops, ham, roasts, bacon, loins
Blood, pancreas, heart	Pigs are the source for nearly forty drugs, pharmaceuticals, and transplant materials including insulin, heart valve transplants, and Heparin (an anticoagulant for humans)
Organs, intestines, and blood	Sausages, kidney pie
Stomach	Haggis
Teeth	Jewelry, buttons

meat tastes different. Highlight the unique qualities of your breed to attract the customers seeking such qualities. For example, pigs with higher fat content are typically attractive to people who enjoy slow cooking and old-fashioned flavors. Diet-conscious individuals seeking lean meat could benefit from some of the breeds known for low fat content. All customers are different, and

Planning for Your Pig's Future

Although most people don't like to think about it, making a plan in the event of an emergency, illness, or death is imperative to the future of your livestock. If you raise animals, you should have a plan in place to take care of them in the event that something happens to you or their primary caregiver.

By making a will that includes your animals, you can designate where they should go and to whom. Make prior arrangements with someone to take your animals or come care for them. Put your instructions in writing, and be certain that everyone involved has a copy of your instructions.

Many purebred herds of livestock have been lost when their owners die and family or neighbors do not know what to do with the animals. Make the decision now to ensure that your animals will be passed along and cared for.

you should advertise your unique pork in a manner that will connect with those seeking the specific attributes your products have to offer.

Selling Breeding Stock

If you raise purebred pigs, get involved with the association for your particular breed. Typically, breed associations offer a breeder's directory, included in the membership fee. Some offer printed as well as Internet listings, both of which can help you find customers. Most people looking for pigs will look closest to their home first.

Many Web sites and animal-oriented magazines offer ad space where you can tell people about your pigs. If cost is an issue, search out the free ones first and start there. You'd be surprised at how many people will find you through these avenues.

Look at Web sites and magazines that offer articles on topics related to what your farm does. For example, if you raise organic pork, advertise in magazines that promote organic and naturally raised products. Avoid magazines that are too far off topic from what you are selling.

The Internet is a very good avenue for reaching lots of people fast. Many sites offer free or nominally priced advertising, including pictures and full-color ads. Many of these sites will help you design an ad for a small fee. These ads are generally more professional looking and will attract interest. Evaluate placing your ad in a priority listing, which will show it as a block ad or as a highlighted ad on some sites. Again, look at sites that are geared toward what you are selling.

When posting an ad, make your pigs stand out from the rest. Tell about the way they are raised or anything special about them. If you can offer delivery or other services, be sure to mention them.

Purebred pigs can be marketed as breeding stock, as specialty meats, and as butcher pigs. By starting with purebreds, you always have the other two markets available. Be an upstanding breeder, and your reputation will soon follow. Remember the section on screening breeders? You want to pass all the requirements you had for the person you bought pigs from.

Have the paperwork ready for any potential purebred pig buyer. Offer a receipt or bill of sale and the registration paperwork necessary for the buyers to register their pigs. If your buyers come to your farm and you are sure they will purchase a pig, have the pig already moved to a convenient location for loading.

Other Options for Selling Pigs

Some farms offer piglets for sale at weaning age with an offer to raise the pig to butcher weight. If you have enough feed stored for the duration of the growing pig's life, you can easily calculate your feed and labor costs for such an endeavor. Counting on this may not be advisable if you are at the mercy of the current grain prices, however.

Alternatively, you can sell weanling pigs to other farmers who wish to raise the pigs themselves. You will probably get only current market price for your pigs, so this may not be a profitable option. If you anticipate a drought or other shortage of feed, or if the market is low for pork, selling your pigs at weaning time may be the best option. You can always sell a few and raise a few to gauge how this system will work for you.

Pigs raised to butcher weight can be marketed through livestock sale houses or directly to butchers. Butchers are often interested in working directly with the farmers to obtain their pigs. By establishing a relationship with a butcher, you are almost guaranteed to sell all your pigs as they become available, especially if you offer pigs that are raised differently from your competitor's pigs or have unique qualities such as fat or flavor. A butcher can tell you approximately how many pigs he can use in a certain time frame, allowing you to plan your production methods and numbers of pigs. Highlight the methods of production you are using to set yourself apart from other producers who are raising pigs in your area.

Conclusion

As you can see, there are a variety of ways in which to market your pigs and pork products. By identifying the market in your area and promoting your products by their virtues, you can realize a decent income through small-scale pig production.

Acknowledgments

Special thanks go to:

- Bret Kortie, co-rancher, co-photographer, and co-conspirator in the saving of rare breeds.
- Dawn and Jim Arkerson, for leading by example and encouraging me to make the world a better place.
- Poppy Tooker and the Slow Food New Orleans gang, for spreading the news of “Eat it to Save it.”
- My friends at the American Livestock Breeds Conservancy, for assistance, encouragement, technical advice, and friendships.
- Mark and Jessica Dibert of the American Mulefoot Hog Association, for encouragement and their tireless dedication.
- Dr. Phil Sponenberg, for teaching me about breeding livestock long before we had ever met.
- Pipestone Veterinary Clinic, for advice, veterinary care, and dedication to rare breeds.

Appendix: Swine Diseases at a Glance

Abscesses

- Abscesses are pus-filled swellings caused by bacterial infections, typically of the *Streptococcus* strain. Additional symptoms include lower feed intake, high fever, constipation, and pain around the abscess.
- Abscesses can be lanced and drained and cleaned with iodine or antiseptic. Lancing may introduce the bacteria to other pigs in the same pen, so isolation is recommended. They are treated with antibiotics.

Atrophic Rhinitis

- This is a bacterial infection that attacks the turbinate bones in the nose, causing them to become damaged and distorted (atrophy). Sneezing and nasal discharge may be the only clinical signs in mild cases. Advanced infections may cause distortion of the nasal bones, whereby the nose twists to one side or develops a hole through the side of the snout.
- Severely affected animals will have difficulty eating and be more susceptible to pneumonia. Atrophic rhinitis is spread through nose-to-nose contact. A vaccine is available to help minimize the symptoms. Atrophic rhinitis is treated with antibiotics and sulfonamides.

Brucellosis

- Brucellosis is caused by the bacterium *Brucella suis.* It is diminishing in occurrence in the United States but may still be found in isolated populations. It is typically spread by venereal infection and passed from the boar to the sow.
- Once infected, the bacteria harbor in the sow's placenta, causing inflammation and abortion. Additional symptoms include swollen testicles (boars), infertility, sterility, lameness, or paralysis of

rear legs. Brucellosis is found in many species, and may be transmissible. Any known carrier should be destroyed.

Coccidiosis/Coccidia

- Coccidiosis is caused by parasites that multiply within host cells, primarily within the intestinal tract. It is usually transmitted by the sow to the piglets through feces contamination. Common in piglets from seven to fifteen days of age, coccidian infestation is only rarely seen in boars or grower pigs. Once infected, the piglets show signs of sloppy diarrhea and can quickly dehydrate and die.
- Treatment can be determined by a laboratory exam of the feces to determine which strain of coccidian is the culprit. Treatments include a coccidiostat, amprolium, or sulphonamides. Immediate treatment is necessary to prevent wasting and death.

Cryptosporidiosis

- Cryptosporidia are parasites that cause diarrhea in piglets eight to twenty-one days of age. Symptoms include watery brown diarrhea and emaciation. Dirty pens and poor hygiene perpetuate cryptosporidia.
- Treatment and prevention principles are the same as for coccidia, including an oo-cide disinfectant (used to clean surfaces in pig housing) that will eliminate coccidia and some viruses, bacteria, and fungi.
- Cryptosporidia can infect humans, rats, mice, and other mammals. This can be particularly dangerous to immune-suppressed humans. Treatment for humans is available.

E. Coli

- The bacteria *E. coli* can quickly invade the intestines of piglets that have not received sufficient colostrum (and thereby did not receive the antibodies required to fight the bacteria). Recently weaned piglets are also susceptible to outbreaks of *E. coli*, as the piglets are no longer drinking the milk that provided protection in the gut.
- *E. coli* infection is characterized by a creamy white diarrhea, and the piglets quickly dehydrate. They may huddle together, shiver, or lie on their sides making a paddling motion. Many treatments are available (antibiotics, sulphonamides). Piglets under 7 days should be treated by mouth. The best defense is a clean environment and highly absorbent bedding material.

Enteritis

- Enteritis is a disease characterized by diarrhea and caused by one of several species of the bacterium *Clostridium.*
- Characterized by an inflammation in the small intestine, enteritis usually affects piglets within 72 hours of birth. Sudden death may occur before you have time to recognize the illness. There are vaccines for some types of *Clostridia*. Special attention to cleanliness, such as washing the sow before

farrowing, may help control the incidence of enteritis.

Erysipelas

- The bacteria *Erysipelas* can be found on most pig farms. It is always present in either the pig or in the environment because it is excreted via saliva, feces, or urine. It is also found in many other species, including birds and sheep. Infected feces are probably the main source of contamination, as the bacteria can survive outside the body for several weeks or longer in light soils.
- Symptoms include diamond-shaped skin lesions, high fever, abortions, mummified piglets, and variable litter size. A vaccine is available. If an outbreak occurs, immediate treatment with fast-acting penicillin is indicated. Moving the pigs to clean areas will help diminish the disease.

Flu

- Swine influenza (SI) is caused by a number of influenza A viruses. The onset is typically rapid and drastic, lasting from twenty-four hours to ten days. It is virtually impossible to keep SI from affecting your pigs at some point. SI can be introduced by people, birds, carrier pigs, and other animals. It is exacerbated by stress, fluctuating temperatures, wet bedding, and poor nutrition.
- Symptoms include lethargy, coughing, fever, abortions, and severely ill pigs that appear to be near death. In fact, most pigs recover without treatment. However, secondary infections and pneumonia can result from the initial viral infection, and affected pigs should be treated with antibiotics.

Glassers Disease

- Known also as *Haemophilus parasuis,* or Hps, this disease is found all over the world, even in herds that have top health records. Hps attacks the smooth surfaces of the joints, coverings of the intestine, the lungs, the heart, and the brain causing pneumonia, heart sac infection, peritonitis, and pleurisy.
- Seen mainly in piglets, symptoms include a short, repeating cough, fever, depression, loss of appetite, lameness, and death. Hps is thought to be brought about by exposure to pneumonia, porcine reproductive and respiratory syndrome (PRRS), flu, drafts, poor environments, and stress. Treatment is effective if administered immediately and should consist of penicillin or oxytetracycline.

Leptospirosis

- Commonly known as lepto, this disease is caused by bacteria that are carried by most mammals. It is transmitted through the urine of animals, from pig to pig and from other wildlife to pigs. Symptoms include loss of appetite, depression, abortions, stillbirths, and weak piglets that die shortly after birth.
- If an outbreak occurs, pigs should be treated with the tetracycline family of antibiotics. Additional manage-

ment practices should include removal of urine, allowing wallows to dry out completely between uses, control of rodents and other livestock coming in contact with the pigs, and vaccinations.

- **Remember**: Leptospira organisms have been found in cattle, pigs, horses, dogs, rodents, and wild animals. Humans become infected through contact with water, food, or soil containing urine from these infected animals. This may happen by swallowing contaminated food or water or through skin contact, especially with mucosal surfaces, such as the eyes or nose, or with broken skin. The disease is not known to spread from person to person.
- Symptoms of leptospirosis in humans include high fever, severe headache, chills, muscle aches, and vomiting and may include jaundice, red eyes, abdominal pain, diarrhea, or a rash. If the disease is not treated, the patient can develop kidney damage, meningitis, liver failure, and respiratory distress.

Lice

- Lice are external parasites found on many farm animals and can be seen by the naked eye. Lice suck blood from their host animals and can be responsible for the transmission of bloodborne illness and anemia.
- Treatment is a simple pour-on delouser or anthelmintic. These treatments are ineffective against the eggs, so a second treatment, given ten days after the first, is required.
- Lice can be transmitted to humans and other livestock.

MMA Syndrome (Mastitis, Metritis, Agalactia)

- MMA is a series of illnesses affecting the post-farrowing sow. Mastitis is an udder infection, characterized by swollen teats, fever, loss of appetite, and an unwillingness to suckle piglets. Metritis is a bacterial infection of the uterus. More likely to occur when assistance was given during farrowing, metritis can also follow mastitis. It is characterized by fever, loss of appetite, and a brown discharge from the vulva. Agalactia is a loss of milk production. These illnesses do not always follow each other or happen as a syndrome, as each can appear on its own for various reasons.
- The key to good milk production and health of the sow during and after farrowing is cleanliness. Clean, dry environments help diminish the effects of bacterial infections. Mastitis and metritis are treated with oxytocin (to promote milk let-down and uterine contractions), antibiotics, and vitamin injections. Adding fiber to the sow's diet and providing the opportunity for exercise also help stimulate milk production. Agalactia that has been brought on by mastitis or metritis can usually be resolved by these measures. Agalactia brought about by old age in the sow cannot be reversed. It is imperative that the herdsman monitor the

piglets. If they are not receiving milk from the sow, then an immediate intervention must be made, as piglets are highly susceptible to starvation during the first few days of life.

Mange

- Mange is a parasitic skin disease caused by mites. Mange is unsightly and extremely uncomfortable for the pigs, as it causes them to scratch and rub the skin to the point of scabs, scrapes, and punctures. In severe cases, pigs may develop an allergic reaction to the mites and have an outbreak of red pimples over the entire body. Spread from pig to pig, mange can be easily introduced to a herd via new animals that have not been treated and quarantined.
- Treatment is through oral, injectable, or pour-on medications. Repeated treatment is necessary. Two doses of a product such as Ivermectin, given ten days apart, should eliminate most cases.
- It is rare for mange or pig mites to be transmitted to humans, as they are host-specific parasites.

Pneumonia

- Characterized by a chronic dry cough, pneumonia is a common bacterial or viral infection in growing pigs and is occasionally seen in gilts with compromised immune systems or current infections. Mortality in naïve herds is high, between 10 and 15 percent, if not rectified immediately. Naïve herds are herds that have never been exposed to or have not been vaccinated for the particular strain of pneumonia. Various strains of pneumonia exist, and they must be differentiated from Actinobacillus Pleuropneumonia (App), enzootic pneumonia, PRRS, flu, and *Salmonella choleraesuis* pneumonia. Additional blood tests can help determine the exact pathogen associated with an outbreak, and a targeted antibiotic may be utilized.
- Other symptoms include fever, rapid breathing, loss of appetite, huddling, and loss of condition. Pneumonia can be compounded by the type and level of pathogens (bacteria and viruses) present in the living quarters, the immunity of the pig at the time of exposure, and the management practices of the herdsman. Additional causes of pneumonia are poor environments with humidity, ammonia, and drafts; overstocking of pens; transient pigs; and poor nutrition.

Pasteurellosis

- Caused by the *Pasteurella multocidia* bacteria, pasteurellosis is commonly indicated by respiratory diseases of pigs between ten and eighteen weeks of age. Pasteurellosis is characterized by severe and sudden pneumonia, with rapid breathing, high fever, and high mortality. Less severe cases may include mild pneumonia, coughing, nasal discharge, and wasting. Thought to be a secondary condition to PRRS, flu, or Enzootic pneumonia, it should be treated the same as for pneumonia.

Porcine Reproductive and Respiratory Syndrome (PRRS)

- PRRS infects all types of herds, including those with high health standards and both indoor and outdoor units, irrespective of size. Severe economic loss can result from an outbreak of PRRS in a herd.
- PRRS was first identified in the midwestern United States and dubbed the mystery swine disease. Caused by an arterivirus (a virus that attacks the macrophages, the white blood cells responsible for ingesting and eliminating bacteria, protozoa, and tumors), there is no known cure for PRRS at this time.
- Symptoms are varied, depending on the age of the pig affected, but they include respiratory and reproductive disorders and can range from loss of appetite to abortions and stillbirths. Fertility issues are also implicated in the disease, with breeding and cycling being delayed. Symptoms can last up to three months and may recur. A vaccine used to control and prevent PRRS has had only limited success in the United States. The best way to avoid PRRS is to purchase only from herds tested to be PRRS free; to quarantine all incoming animals to avoid exposure to existing pigs; to maintain dry, clean facilities that are not in continual use by pigs; and to minimize outside visitors to the farm.

Pseudorabies

- A herpes virus found in pigs, pseudorabies can remain dormant in the pig's nerves for long periods and then become reactivated. Once introduced to a herd, the virus remains and affects fertility to varying degrees. Periods of stress may activate the disease, which can be transmitted through nose-to-nose contact and via air. Other sources of contaminants can be feral animals, people, vehicles, semen, and contaminated carcasses.
- Prominent symptoms involve the nervous and respiratory systems, such as coughing, sneezing, incoordination, reproductive failure, abortions, stillbirth and mummified piglets, and piglet death. A vaccine is available but should be considered with extreme caution. Once a pig is vaccinated for pseudorabies, it will test positive for the virus. This may prevent you from selling animals as breeding stock in some states. Pseudorabies can be transmitted from pigs to cattle, horses, dogs, and cats, which develop nervous signs, itching, and rapidly die. It is not known to cause disease in people.

Salmonellosis

- Salmonellosis is caused by the bacteria *Salmonella*, and the disease comes in many forms. *S. choleraesuis, S. typhimurium,* and *S. derby* are the most common acute forms seen in pigs. If a pig is infected by a large number of *S. choleraesuis* organisms, it is likely to develop severe disease, starting with a septicemia (blood infection) and followed by severe pneumonia and enteritis. Mild exposure to the disease may be asymptomatic, as it can lie dormant within the intestines of the pig. Afterward, the *S. choleraesuis* may migrate into a variety of tissues, including the central nervous system and joints, resulting in meningitis and arthritis. *S. typhimurium* and *S. derby* may also cause septicemia and pneumonia, but in most pigs enteritis is the only persistent manifestation. Subacute forms are characterized by foul-smelling diarrhea.

- It is imperative that professional help be sought for the diagnosis and treatment of Salmonellosis because there are so many prevalent strains. Laboratory analysis of fecal samples can determine the type of bacteria and indicate the appropriate treatment. Maintain the highest cleanliness standards, minimize rodent infestation of feed, do not mix groups of pigs, remove feces frequently, and disinfect areas as pigs are removed.
- It is important to remember that salmonella organisms are transmissible to people and are one of the most common causes of food poisoning. Ensure that everyone working with the pigs adopts a high standard of personal hygiene to minimize the risk of infection.

Toxoplasmosis

- Toxoplasmosis is a protozoan that affects humans and animals. Cats are the primary source of toxoplasma, and they shed the disease in the feces. Pigs may become infected through eating infected cat droppings, rodents, or other pigs. The organisms form cysts within the pig's muscles and organs. When humans ingest the meat, the cysts can develop into mature parasites.
- Clinical signs of toxoplasmosis in pigs are uncommon. Confirm suspected signs with a veterinarian. Toxoplasmosis is typically treated with sulfonamides.
- In humans, pregnant women are most at risk from a toxoplasma infection, as it may cause abortion and birth defects. Infection can be confirmed with a blood test. Most humans never show clinical signs of infestation.

Transmissible Gastroenteritis (TGE)

- TGE is a highly contagious disease caused by the coronavirus. The older a pig, the better the chance for recovery. Immediate response to TGE is critical. Piglets must be monitored and provided electrolytes, neomycin, extra bedding, and an extra heat source. A once common practice was to infect an entire herd to produce immunity in the sows.
- Symptoms in small pigs include diarrhea, vomiting, dehydration, and death. There is a vaccine available. TGE can be spread by the herdsman, and care should be taken to clean boots between pens, to keep dogs and birds away from your pigs, and to not visit other farms that raise pigs. All incoming stock should be tested for TGE.

Glossary

Abattoir—Slaughterhouse

Anthelmintic—A treatment to rid the body of worms.

Barrow—A castrated male pig. Most commonly used as a meat pig.

Blind Teat—Small or functionless teat. Some producers remove the teat to prevent a piglet from attaching to a teat that will not provide nourishment.

Boar—An intact male raised for breeding purposes.

Body Capacity—Used to describe the shape of the hog in relation to health, feed conversion, growth, and carcass quality potential.

Bone—Used to describe the width of head and diameter of bone in the legs and jaw. Important in gauging the durability and capacity of the animal to carry flesh.

Bulk—Fiber added to the diet or mixed into a grain ration.

Butcher Hog—Also known as Market Hog, weighing between 220–260 pounds, approximately six months of age, raised specifically for the slaughter market.

Carcass—The dressed out (butchered) body of a hog.

Chitterlings—The small intestines of a pig.

Colostrum—The first milk after farrowing, full of antibodies, fat, and nutrition whereby the sow passes on valuable immunity to the piglets. Produced for approximately seventy-two hours.

Cracklings—Bits of browned meat or skin that result from rendering lard—also known as pork rinds.

Creep Feed—Ground or pellet feed offered to piglets as starter ration, high in sugar and milk protein. Walk-in feeders that surround the creep feed and keep the sow from invading are called creep feeders.

Crossbreeding—Mating of hogs of different breeds to maximize the qualities of both.

Culling—Removal of animals from the herd, usually for performance, disposition, or health reasons.

Drove—Herd or group.

Electrolytes—Mineral salts used by the body to help absorb liquids, used to mend problems associated with dehydration, fever, diarrhea, and prolonged illness.

Embryo—A piglet in early stages of growth in the sow's uterus.

Estrus—The period in which a sow will allow service from the boar; lasts one to three days.

F1—The first generation of offspring as the result of mating between two purebreds of differing breeds. F1's typically display hybrid vigor.

Farrowing—Giving birth.

Farrow to Finish—Raising a piglet from birth to butchering.

Feed Efficiency—The amount of feed an animal must eat to gain one pound of flesh.

Feeder Pig—A young pig, usually just after weaning, that is produced by one farm, but raised by another to the butchering weight. Or a pig being reared for pork.

Finish Hog—A hog that has been fattened and is ready for market or slaughter.

Finishing—Feeding out a hog to slaughter weight.

Fitting—Readying a hog for showing or exhibition.

Following—Practice of letting pigs graze or clean up after cattle to glean any unused grains or nutrients from the manure. Also used to turn compost or loosen up packed feedlot waste.

Frame—The skeletal system of the hog. A large frame indicates substance, stability, and durability.

Full Feed—Allowing the hog to consume all the feed it wants, usually practiced in grower/finisher operations.

Genotype—Represents an organism's exact genetic makeup; that is, the full hereditary information of an organism. The genotype can indicate the size, performance, and disposition of an animal.

Gestation—The timeframe of the pregnancy. Typically last 3/3/3; that is, 3 months, 3 weeks, and 3 days on average.

Gilt—Female pig that has yet to bear young.

Grading—Sorting pigs according to their quality of carcass.

Hand Breeding—Breeding directly supervised or assisted by the herdsman. Sow and boar are generally separated immediately after intercourse.

Heterosis—Hybrid vigor, the increased strength of superior qualities arising from the crossbreeding of genetically different animals.

Hog—A pig who has grown to at least 120 pounds.

Hogging Down—Allowing pigs to harvest or clean up crop fields.

Hot Wire—An electrified wire or fence.

Hurdle—Solid board or gate used to direct pigs when moving. See Pig Boards.

Inbreeding—Mating two closely related individuals to concentrate or maximize certain traits. This can bring out the good and bad traits. A popular farmer's quote is "linebreeding is when it works, inbreeding is when it doesn't."

In-Pig—A pregnant sow.

Lactation—Time when the sow is producing milk and nursing piglets.

Lard—Rendered hog fat.

Libido—Sexual interest aroused in animals when they are receptive to breeding.

Limit Feeding—Restricting the amount of feed an animal is given, usually practiced with sows and boars to avoid obesity and breeding difficulties.

Line—Specific bloodlines tracing back to individuals or family groups.

Line Breeding—Repeat breeding to a single individual to concentrate the desirable genetics (that is, parent to offspring, sibling to sibling).

Litter—The offspring from a single farrowing. United States national average is seven piglets per litter.

Market Hog—Same as butcher hog.

Mash—Ground feed usually fed to piglets and infirm animals.

Meal—Finely ground feed.

Mummy—Piglets born dead, typically having no hair and being only partially developed. They died too late in the pregnancy to be reabsorbed by the sow.

Natural Immunity—The immunity that animals build up over time as the result of exposure to disease-causing bacteria or viruses. Immunity is passed through colostrum as well.

Needle Teeth—Also called eye, canine, or wolf teeth, two teeth on either side of the upper and lower jaws. These are very sharp, and can cause injury to the sow's teats or to littermates when fighting over teat order. Typically clipped right after birth.

Offal—The internal organs of a butchered animal, although a frequently used term for the waste parts of the animal.

Open—Term used to describe a female pig who did not conceive at breeding or one who aborted or absorbed a pregnancy.

Overlaying—Sow crushes and kills her piglets when laying down.

Oxytocin—Drug used to cause uterine contractions or stimulate milk production.

Pathogen—A disease-causing agent.

Phenotype—The physical characteristics displayed in an individual such as height, weight, hair color, and so on.

Pig—A very young swine.

Pig Boards—30- to 40-inch-high boards with handle holes cut along the top, used to place alongside or in front of pigs as a guide while moving them.

Placenta—The sac within a sow that houses the piglets until farrowing. Expelled after all piglets are born.

Premix—Vitamin and mineral supplement added to ground feed to make a completely balanced ration.

Prepotent—Tendency of a boar or sow to pass on specific genetic traits.

Primal Cuts—Large portions of the butchered carcass, such as whole sides.

Probiotics—Beneficial organisms, usually lactobacillus, used to repopulate the flora in a sick or recuperating pig's gut, or to stimulate nutrient absorption.

Prolificacy—Number of offspring produced by any sow or boar.

Purebred—Animals bred of a specific breed for many generations, usually with pedigrees to verify such lineage.

Render—Heating fat to remove impurities and moisture prior to storage.

Ring—To place a metal ring or rings into the snout of a pig to prevent rooting.

Rotational Grazing—Moving animals from one to another consecutive pastures to maximize feed utilization, minimize parasite infestation, or to revitalize weedy or ailing pastures.

Scours—Diarrhea

Scrotum—The sac on a boar that holds the testicles, it contracts or expands as needed to regulate temperature in the testicles, and preserve sperm.

Service—When the boar has bred the sow or gilt.

Settle—When the sow or gilt becomes pregnant, or a pregnancy "sticks."

Shoat—Young pig from weaning to 120 pounds.

Single tree—The bar used to attach harness equipment to a cart horse.

Snare—A device that fits around the snout for the purposes of restraining the pig.

Soft Pork—Pork that remains soft, flabby, or oily even after chilled, usually produced by improper feeding. Undesirable for meat production.

Sow—A female pig that has previously farrowed.

Stag—A boar who has been castrated after being used as a stud boar.

Standing Heat—Period in gilt's or sow's heat cycle when she will stand still and rigid waiting for the boar to mount.

Swine—A generic term for any pig.

Terminal—A boar used to create crossbred pigs (F1) but not bred to subsequent generations, as the offspring are destined for butchering.

Top Dress—To manually add supplements to a hog's ration.

Underline—Usually referred to with sows, as in evaluating the teat placement and udder line for adequate milk production. Also may be used to evaluate sheath and penis in boars.

Vulva—The external portion of the female genitals. Changes in vulva can indicate heat or preparation for farrowing.

Weanling—Piglet recently removed from the sow, usually between 25 and 40 pounds.

Registries

American Berkshire Association
PO Box 2436
1769 U.S. 52 West
West Lafayette, IN 47906
(765) 497-3618

American Landrace Association
PO Box 2417
West Lafayette, IN 47906
(765) 463-3594

American Mulefoot Hog Association and Registry
18995 V Drive S.
Tekonsha, MI, 49092
(517) 767-4729
http://www.mulefootpigs.tripod.com

American Yorkshire Club
PO Box 2417
West Lafayette, IN 47996
(317) 463-3593
Fax (317) 497-2959

British Saddleback Breeders Club
Dryft Cottage
South Cerney
Cirencester, Gloucestershire,
GL7 5UB
01285 860229
http://www.saddlebacks.org.uk
mail@saddlebacks.org.uk

Certified Pedigreed Swine
PO Box 9758
Peoria, IL 61612
(309) 691-0151

Gloucestershire Old Spots Pigs of America
1275N 1900E
Dewitt, IL 61735
(309) 928-3987
http://www.gosamerica.org

Hampshire Swine Registry
1803 W. Detweiler Dr.

Peoria, IL 61615
(309) 692-1571
Fax (309) 691-8178

Large White (see American Yorkshire Club)

Middle White Breeders Club
Miranda Squire
Benson Lodge, 50 Old Slade Lane
Iver, Buckinghamshire, SL0 9DR
01753 654166
miranda@middlewhites.freeserve.co.uk

National Hereford Hog Record Association
22405 480th Ave.
Flandreau, SD 57028
(605) 997-2116

National Spotted Swine Record
6320 N. Sheridan Road
Peoria, IL 61614
(309) 691-0151
Fax (309) 691-0168

National Swine Registry
1769 US 52 West
P.O. Box 2417
West Lafayette, IN 47996-2417
(Registry for Hampshire, Landrace, Duroc, and Yorkshire pigs)

North American Large Black Pig Registry
Still Meadow Farm
740 Lower Myrick Road
Laurel, MS 39440
(601) 426-2264

North American Potbellied Pig Association
2016 64th St. Court East
Bradenton, FL
(941) 746-7339

Ossabaw Island Hog Registry
W 1205 Nenno Rd.
Mayville, WI 53050
(920) 387-5991
http://www.ossabaw.org

Poland China (see National Swine Registry)

Red Wattle Hog Association
21901 Mayday Road
Barnes, Kansas 66933
(785) 944-3574

Oxford, Sandy and Black Pig Society
Lower Coombe Farm
Blandford Road
Coombe Bissett
Salisbury, Wilts. SP5 4LJ
01722 718263
http://www.oxfordsandypigs.co.uk

Tamworth Swine Association
200 Centary Rd.
Winchester, OH 45697
(765) 653-4913

United Duroc Swine Registry (see National Swine Registry)

Rare Breed and Animal Preservation Societies

American Livestock Breeds Conservancy
PO Box 477

Pittsboro, NC 27312
(919) 542-5704
http://www.albc-usa.org

FERME (French for Association for the Promotion of Endangered Domestic Breeds)
Le Bourg
42600 Grézieux-le-Fromental
http://www.chez.com/ferm

New Zealand Rare Breeds
170 Tuahiwi Road
R D 1, Kaiapoi 7691
New Zealand
http://www.rarebreeds.co.nz

Rare Breeds Canada
1-341 Clarkson Road
RR1, Castleton ON, K0K 1M0
Phone/Fax: (905) 344-7768
http://www.rarebreedscanada.ca

Rare Breeds International
Aristotle University of Thessaloniki
Animal Genetics and Breeding
54124 Thessaloniki, Greece
+302310998683/87
http://www.rbi.it

Rare Breeds Survival Trust (RBST)
Stoneleigh Park
Nr Kenilworth
Warwickshire
CV8 2LG
http://www.rbst.org.uk

Rare Breeds Trust of Australia
PO Box 220
Heathcote, Victoria, Australia 3523
03 54-333236
http://www.rbta.org

Helpful Web Sites

http://www.ams.usda.gov/nop/indexIE.htm
National Organic Program home page. Full of information on becoming a certified organic producer, including lists of certifying organizations grouped by state.

http://www.mosesorganic.org/factsheets/index.htm
Midwest Organic and Sustainable Education Service. A series of fact sheets about the organic certification process.

http://www.attra.org/organic.html
Appropriate Technology Transfer for Rural Areas (ATTRA). The source of information about all kinds of sustainable and organic agriculture topics. You will also find links to other publications about organic certification and transition here.

http://www.templegrandin.com
Information on stress-free handling of livestock, facilities planning, and intuitive animal husbandry.

http://www.maveric9.com
Heritage Ranch Co. home page, offering information on rare hog breeds.

http://www.aspca.org
ASPCA home page, offering information on toxic plants.

http://www.thepigsite.com
Comprehensive site on health and management of hogs, plus American and foreign pig issues.

http://www.askthemeatman.com
A very helpful site on butchering tips and meat curing.

Publications and Periodicals

Countryside Magazine
145 Industrial Dr.
Medford, WI 54451

Hogs Your Way
University of Minnesota Extension Service
Distribution Center
405 Coffey Hall
1420 Eckles Avenue
St. Paul, MN 55108-6068

Kansas State University
Agricultural Experiment Station and Co-Op Extension
http://www.Oznet.k-state.edu

Mother Earth News
1503 SW 42nd St.
Topeka, KS 66609

Rare Breeds Journal
PO Box 66
Crawford, NE 69339
(308) 665-1431
rarebreed@bbc.net

Small Farm Today
3903 W. Ridge Trail Road
Clark, MO 65243-9525
(573) 687-3525
smallfarm@socket.net

Books

Caras, Roger. *A Perfect Harmony.* Simon and Schuster, 1996.
Interesting read on the relationship between man and domestic animals through the ages.

Dohner, Janet. *The Encyclopedia of Historic and Endangered Livestock.* Yale University, 2001. History, characteristics, and qualities of 138 endangered breeds.

English, Peter. *The Sow—Improving Her Efficiency.* Farming Press LTD, 1982.
Technical but very informative book on maintaining sows for optimum performance.

Hedgepeth, William. *The Hog Book.* Doubleday, 1978.
An interesting read about the potential of hogs in history, literature, love, sports, show business, art, poetry, and where pigs may go in the future.

Henderson, Fergus. *The Whole Beast, Nose to Tail Eating.* Ecco Publishing, 2004.

Recipes from the owner of St. John Restaurant, making delicious meals with everything but the squeal.

Kaminsky, Peter. *Pig Perfect.* Hyperion, 2005.
Fascinating book about one man's quest for the ultimate ham.

Lust, John. *The Herb Book.* Bantam Books, 1974.
A complete guide to identifying and using plants, herbals, and medicinals, including toxic plants.

Porter, Valerie. *Practical Rare Breeds.* Pelham Books, 1987.
Practical and useful book on the husbandry methods for rare livestock breeds.

Ruhlman, Michael and Brian Polcyn. *Charcuterie.* W. W. Norton, 2005.
Gorgeous cookbook on salting, smoking, curing, and cooking meats.

Smedley, John. *Home Pork Production.* Orange Judd, 1943.
Management of pigs to secure pleasure and profit.

Watson, Lyall. *The Whole Hog.* Profile Books, 2004.
A biologist's tribute to the *Suidae* family, showing pigs as intelligent, gregarious, and practical.

SUPPLIES

http://www.countryhorizons.net
Pig supplies for beginner and experienced hog persons.

http://www.enasco.com
Large selection of supplies for farrowing to finishing.

http://www.hogslat.com
Farrowing supplies, pens, and equipment for hogs.

http://www.mytscstore.com
Farming supply store, check local listings also.

http://www.premier1supplies.com
All your fencing needs, instructions, and net fence.

http://www.vittetoe.com
Many automated supplies and time savers for hog owners.

Index

G

H

I

K

Q

R

S

T

U

V

W

Y

Z

About the Author

Arie McFarlen, PhD, is a co-owner of Maveric Heritage Ranch Co. in South Dakota, a ranch dedicated to saving and promoting endangered livestock breeds. She is the author of several articles on endangered breeds and their care featured in *Hobby Farms* magazine, *Rare Breeds Journal*, and *Small Farm Journal.*